MILITARY WRITINGS

By Leon Trotsky

PATHFINDER PRESS, INC.
NEW YORK

CONTENTS

PREFACE

Unlike subsequent anticapitalist revolutions in countries such as Yugoslavia, China and Cuba, the Russian revolution of 1917 came to power almost bloodlessly; only after that was it compelled to fight a civil war. Having state power was, of course, an enormous advantage for the Bolsheviks in this war against capitalist counterrevolution and imperialist intervention. On the other hand, the country had been devastated by four years of World War I, the economy lay in ruins, and the people were exhausted. In addition, the Bolsheviks had to start virtually from scratch in assembling an army capable of defending the new workers' state in a conflict that for three years raged back and forth over most of the former Russian empire.

Leon Trotsky (1879-1940) is best remembered today as Lenin's co-worker in the revolution of 1917 and as the leader of the struggle for internationalism and workers' democracy waged from 1923 against the Soviet bureaucracy headed by Stalin. But he served the revolutionary cause in many other ways. His first post in the new Soviet government was that of Commissar of Foreign Affairs; the assignment he was given immediately after the 1917 insurrection was to try at the negotiations held in Brest-Litovsk to bring about the end of World War I or, failing that, to extricate Russia from the war. After the Germans imposed a brutal peace on Russia in March, 1918, the civil war to overthrow the Soviet government began in real earnest. Although Trotsky had had no formal military training, he was then appointed Commissar of War, a position he filled not only throughout the civil war but until 1925, after Lenin's death. As chairman of the Supreme Military Council, and later of the Revolutionary Military Council, he organized the Red Army. When the civil war ended, he helped to reorganize the Red Army.

How contemporary Bolshevik opinion evaluated Trotsky's

military achievement is accurately reflected in "The Organizer of Victory," an article written for *Pravda* by Karl Radek in 1923 while Lenin was still alive, which is reprinted here as the introduction to this book. Radek, a German revolutionary of Polish origin who moved to Russia in 1917, was at the time of this article a leader of the Communist International. During the civil war itself Lenin, trying to win Maxim Gorky to support of the Soviet government, had said about Trotsky: "Show me another man able to organize almost a model army within a single year and win the respect of military experts. We have such a man. We have everything. And we shall work wonders."

Being a writer as well as a fighter, Trotsky not only organized and led the Red Army but wrote about it and the military problems facing the first workers' state. It would take many volumes to reprint all of his military writings. *How the Revolution Armed Itself*, published by the Supreme Military Council in three volumes between 1923 and 1925, is over 1700 pages without the notes. Later and briefer accounts of the civil-war period will be found in Trotsky's *My Life* (1930) and in his biography of Stalin (unfinished when he was assassinated in 1940); he also hoped to write a separate book about it. Isaac Deutscher's *The Prophet Armed* has useful chapters on the civil war and a special eight-page note on Trotsky's military writings.

In his writings Trotsky took up all of the many military-political-technical questions posed to the Soviet leadership during the first five years of the Red Army — took them up, argued and fought over them. For they were all "controversial." There were no precedents or traditions to guide these pioneers; the top Bolshevik leaders were often divided over practical measures as well as theoretical problems; and the fate of the revolution often hinged on their answers.

Most of the present collection does not deal, except incompletely or in passing, with the problems debated during the civil war — how the Red Army was built from a few thousand volunteers to a force embracing five million people; how discipline was achieved; whether or not to use former czarist army officers and military specialists, and how to control them; whether to use conscription; whether to have a centralized army with a single command, or guerrilla forces each acting on their own; whether it was more democratic and more efficient to have the officers elected by the troops or appointed by a government elected and controlled by a majority of the people, etc.

Instead, the bulk of this book is focused on disputes that arose after the civil war, during a six-month period at the end of 1921 and the start of 1922. This was shortly after the Communist International concluded, despite some ultraleft opposition at its third world congress, that European and world capitalism had regained the initiative lost at the end of the war, and that the revolutionary movement faced difficult, up-hill and defensive tasks.

The military-political questions that came to the fore at this time concerned the lessons of the civil war and which of them were applicable for other countries where workers would take power and have to create their own army; how the army of a workers' state differs from the army of a capitalist state, whether it has or should have a unique military doctrine, and whether a special kind of military strategy flows from the class character of the state it serves.

Trotsky's views on these questions were given in talks at conferences of the Scientific Military Society attached to the Military Academy of the Red Army and at a meeting of mil-itary delegates to a Communist Party congress, in a pamphlet published by the Supreme Military Council and, two years later, in a review of a book by Frederick Engels. Their first American publication in book form will surely interest students of military affairs.

But that is not the main reason for their present publication, nor is that the audience for whom it will be most useful. The chief value of this collection is its exposition and defense of Marxism — not as an abstraction, but as a concrete applica-tion of the dialectical materialist method of analysis to the superstructure of society (in this case a rather remote part of the superstructure: military activities and relations).

Learning what Marxism is and what it can do is often enhanced by learning simultaneously what it is not and what it cannot do. These writings on "military doctrine" fall chronolog-ically between Lenin's *"Left Wing" Communism: An Infantile Disorder* (1920) and Trotsky's own *Literature and Revolution* (1923) and are on an educational par with them. Lenin used the simple-minded notions of the ultralefts on parliamentarism, boycotts, compromise, retreats, etc., to illustrate the fertile, supple and realistic character of genuinely revolutionary strat-egy and tactics. *Literature and Revolution* similarly enriched the arsenal of Marxism by counterposing to the arguments

of the advocates of "proletarian" literature and art, allegedly
flowing from the class nature of the workers' state, a creative
evaluation of culture and society and their reciprocal relations.

"Our superiority [over our enemies]," said Trotsky in the
discussion of military doctrine, "lies in possessing the irreplaceable scientific method of orientation — Marxism. It is
the most powerful and at the same time subtle instrument —
to use it is not as easy as shelling peas. One must learn how
to operate with it." His opponents in this discussion had not
yet learned how; some never did. Most of them were dedicated
revolutionaries, but relatively new to the Marxist method,
and many suffered in addition from the sickness of ultraleftism.

The military doctrinaires, looking back at the civil war in
Russia, tried to deduce eternal truths and universal principles,
along with strategy and tactics, from experiences that had
fitted certain times, places and conditions. They employed
revolutionary terminology but since, like more recent radicals,
they were contemptuous toward "old knowledge" (the cultural
heritage) merely because it was old, and were attracted to
Marxism because they saw in it a shortcut to ready-made
prescriptions, they succeeded only in muddling things and
producing dogmatic postulates derived from "a rigid methodism," which could only lead to catastrophe if relied on as a
guide for the revolutionary movement in the Soviet Union or
anywhere else.

Trotsky took up and answered each of their military arguments and proposals (their rejection of positional warfare,
their idealization of the offensive under all circumstances,
their explanation for the prevalence of maneuverability in the
civil war, etc.), but since he was defending the Marxist method
as much as a particular military policy, he went far beyond
them. Marxism is a science, an indispensable science for revolutionaries, he explained, but it is a certain kind of science and
not a substitute for other sciences. It can explain the history
and development of trade, or even of chess, but you cannot
trade or create a new bookkeeping system "according to Marx,"
and you cannot learn how to play chess in a Marxist way.
"Behind the attempts to proclaim Marxism as the method of
all sciences and arts there frequently lurks a stubborn refusal
to enter new fields." Thinking you have a master key that opens
all doors and locks is easier than studying bookkeeping,
military affairs, etc. Trotsky also insisted that war is not a
science but an art, even though it is, like architecture, an art

based on and requiring expertness in several sciences.

Readers seeking light on strategical and tactical problems of a political rather than a military character will undoubtedly find material for consideration in Trotsky's discussion of maneuvers and defensive formulations as methods for organizing and winning mass support. Above all, however, these writings are a treasury of dialectical materialism in the hands of a man who had truly learned to use and operate with it. There are few examples in Marxist or any other literature as lucid, profound and fruitful as Trotsky's exposition about the interdependence and interpenetration of the offensive and the defensive, of attack and retreat. In this respect the present book ranks with Trotsky's last and most mature writings, which have been collected under the title of *In Defense of Marxism.*

A note of caution to the reader in the post-Cuban revolution era: Trotsky's denunciations of "guerrillaism" in this book should not be misread to imply opposition to guerrilla warfare in general. What he was opposing here was guerrilla war *after the revolution had been made and a workers' state and its army had been created,* and the tendency of some regional commanders in the early part of the civil war to carry out orders they liked while disregarding those they disliked. In the current debate over guerrilla war, nothing in these writings is pertinent other than the rather obvious point that each national situation must be examined concretely and that actions suitable in one situation should not be mechanically transposed to other situations. On the other hand, there is a marked difference between Trotsky's attitude to the cultural heritage of the past and the need to assimilate its positive aspects before being able to transcend them, and the attitude fostered in China during the "proletarian cultural revolution."

The two appendices in this collection concern another aspect of Trotsky's military writings — the growth of bureaucratism in the Soviet army *and elsewhere,* against which, from Lenin's death to his own, Trotsky was to fight with increasing vigor as the bureaucracy became more uncontrolled, privileged, conservative and hostile to revolutionary internationalism. The first of these, dated only fourteen months after the revolution of 1917, shows how early Trotsky saw this danger manifesting itself and warned against it; the second shows him preparing to challenge the bureaucracy a few weeks before Lenin died.

*   *   *

The first four chapters of this book make up the section called "Military-Technical Problems" in the 1925 edition of *How the Revolution Armed Itself*, Vol. III, Book 2. Translated and edited by John G. Wright, they were first published in English in *Fourth International*, in its issues from December, 1943, through July, 1944.

"'Unified Military Doctrine'" consists of opening remarks and a summary speech before the Scientific Military Society on November 21, 1921, the first anniversary of the founding of the society.

"Military Doctrine or Pseudo-Military Doctrinairism" is a pamphlet completed December 5, 1921, and published by the Supreme Military Council.

"Our Current Basic Military Tasks" consists of a report and a summary speech at a conference of military delegates to the eleventh congress of the Communist Party on April 1, 1922.

"Marxism and Military Knowledge" contains opening remarks and a summary speech at a session of the Scientific Military Society on May 8, 1922.

The fifth chapter, here given the title, "Marxism and Military Warfare," was written March 19, 1924, as a review, "A New Book by F. Engels." It was printed in *Pravda* March 28, 1924, and was first published in English in *The New International*, May, 1944.

The first appendix, "Scientifically—or 'Somehow'?" was a letter written on January 10, 1919, when Trotsky was traveling between the fronts of the civil war ("in the train, between Tambov and Balashov"). It was printed in the Soviet journal, *Military Affairs*, February 23, 1919, and reprinted in *How the Revolution Armed Itself*, Vol. I. The first complete English translation, by Brian Pearce, appeared in *Labour Review*, July-August, 1959.

The second appendix, "Functionarism in the Army and Elsewhere," dated December 3, 1923, first appeared in *Pravda* the next day, and then as an appendix in Trotsky's book, *The New Course*. The translation from that book is by Max Shachtman.

George Breitman
1969

INTRODUCTION

The Organizer of Victory
By Karl Radek

History has prepared our party for various tasks. However defective our state machinery or our economic activity may be, still the whole past of the party has psychologically prepared it for the work of creating a new order of economy and a new state apparatus. History has even prepared us for diplomacy. It is scarcely necessary to mention that world politics have always occupied the minds of Marxists. But it was the endless negotiations with the Mensheviki that perfected our diplomatic technique; and it was during these old struggles that Comrade Chicherin * learned to draw up diplomatic notes. We are just beginning to learn the miracle of economics. Our state machinery creaks and groans. In one thing, however, we have been eminently successful—in our Red Army. Its creator, its central will, is Comrade L. D. Trotsky.

Old General Moltke, the creator of the German army, often spoke of the danger that the pen of the diplomats might spoil the work of the soldier's sabre. Warriors the world over, though there were classical authors among them, have always opposed the pen to the sword. The history of the proletarian revolution shows how the pen may be re-forged into a sword. Trotsky is one of the best writers of world socialism, but these literary advantages did not prevent him from becoming the leader, the leading organizer of the first proletarian army. The pen of the best publicist of the revolution was re-forged into a sword.

------

* Gregory Chicherin was Soviet Commissar of Foreign Affairs at that time (1923).

Marxist Military Literature Was Scant

The literature of scientific socialism helped Comrade Trotsky but little in solving the problems which confronted the party when it was threatened by world imperialism. If we look through the whole of pre-war socialist literature, we find—with the exception of a few little-known works by Engels, some chapters in his *Anti-Duehring* devoted to the development of strategy, and some chapters in Mehring's excellent book on Lessing, devoted to the war activity of Frederick the Great— only four works on military subjects: August Bebel's pamphlet on militia, Gaston Moch's book on militia, the two volumes of war history by Schulz, and the book by Jaures, devoted to the propaganda of the idea of the militia in France. With the exception of the books of Schulz and Jaures, which possess high value, everything which socialist literature has published on military subjects since Engels' death has been bad dilettantism. But even these works by Schulz and Jaures afforded no reply to the questions with which the Russian Revolution was confronted. Schulz's book surveyed the development of the forms of strategy and military organizations for many centuries back. It was an attempt at the application of the Marxian methods of historical research, and closed with the Napoleonic period. Jaures' book—full of brilliance and sparkle—shows his complete familiarity with the problems of military organization, but suffers from the fundamental fault that this gifted representative of reformism was anxious to make of the capitalist army an instrument of national defense, and to release it from the function of defending the class interests of the bourgeoisie. He therefore failed to grasp the tendency of development of militarism, and carried the idea of democracy *ad absurdum* in the question of war, into the question of the army.

Origin of the Concept of the Red Army

I do not know to what extent Comrade Trotsky occupied himself before the war with questions of military knowledge. I believe that he did not gain his gifted insight into these questions from books, but received his impetus in this direction at the time when he was acting as correspondent in the Balkan war, this final rehearsal of the great war. It is probable that he deepened his knowledge of war technique and of the mechanism of the army, during his sojourn in France (during the war), from where he sent his brilliant war sketches to the Kiev

Mysli. It may be seen from this work how magnificently he grasped the spirit of the army. The Marxist Trotsky saw not only the external discipline of the army, the cannon, the technique. He saw the living human beings who serve the instruments of war, he saw the sprawling charge on the field of battle.

Trotsky is the author of the first pamphlet giving a detailed analysis of the causes of the decay of the International. Even in face of this great decay Trotsky did not lose his faith in the future of socialism; on the contrary, he was profoundly convinced that all those qualities which the bourgeoisie endeavors to cultivate in the uniformed proletariat, for the purpose of securing its own victory, would soon turn against the bourgeoisie, and serve not only as the foundation of the revolution, but also of revolutionary armies. One of the most remarkable documents of his comprehension of the class structure of the army, and of the spirit of the army, is the speech which he made—I believe at the first Soviet Congress and in the Petrograd Workers' and Soldiers' Council—on Kerensky's July offensive. In this speech Trotsky predicted the collapse of the offensive, not only on technical military grounds, but on the basis of the political analysis of the condition of the army.

"You"—and here he addressed himself to the Mensheviki and the SR's—"demand from the government a revision of the aims of the war. In doing so you tell the army that the old aims, in whose name Czarism and the bourgeoisie demanded unheard-of sacrifices, did not correspond to the interests of the Russian peasantry and Russian proletariat. You have not attained a revision of the aims of the war. You have created nothing to replace the Czar and the fatherland, and yet you demand of the army that it shed its blood for this *nothing*. We cannot fight for *nothing*, and your adventure will end in collapse."

The secret of Trotsky's greatness as organizer of the Red Army lies in this attitude of his towards the question.

All great military writers emphasize the tremendously decisive significance of the moral factor in war. One half of Clausewitz's great book is devoted to this question, and the whole of our victory in the civil war is due to the circumstance that Trotsky knew how to apply this knowledge of the significance of the moral factor in war to our reality. When the old Czarist army went to pieces, the minister of war of the Kerenski government, Verkhovsky, proposed that the older military classes be discharged, the military authorities behind the front

partly reduced, and the army reorganized by the introduction
of fresh young elements. When we seized power, and the
trenches emptied, many of us made the same proposition. But
this idea was the purest Utopia. It was impossible to replace
the fleeing Czarist army with fresh forces. These two waves
would have crossed and divided each other. The old army had
to be completely dissolved; the new army could only be built
up on the alarm sent out by Soviet Russia to the workers and
peasants, to defend the conquests of the revolution.

When, in April 1918, the best Czarist officers who remained
in the army after our victory met together for the purpose of
working out, in conjunction with our comrades and some
military representatives of the Allies, the plan of organization
for the army, Trotsky listened to their plans for several days
—I have a clear recollection of this scene—in silence. These
were the plans of people who did not comprehend the upheaval
going on before their eyes. Every one of them replied to the
question of how an army was to be organized on the old pattern.
They did not grasp the metamorphosis wrought in the human
material upon which the army is based. How the war experts
laughed at the first voluntary troops organized by Comrade
Trotsky in his capacity as Commissar of War! Old Borisov, one
of the best Russian military writers, assured those Communists
with whom he was obliged to come in contact, time and again,
that nothing would come of this undertaking, that the army
could only be built up on the basis of general conscription,
and maintained by iron discipline. He did not grasp that the
volunteer troops were the secure foundation pillars upon which
the structure was to be erected, and that the masses of peasants
and workers could not possibly be rallied around the flag of
war again unless the broad masses were confronted by deadly
danger. Without believing for a single moment that the volunteer army could save Russia, Trotsky organized it as an apparatus which he required for the creation of a new army.

Utilizing the Bourgeois Specialists

But Trotsky's organizing genius, and his boldness of thought
are even more clearly expressed in his courageous determination
to utilize the war specialists for creating the army. Every good
Marxist is fully aware that in building up a good economic
apparatus we still require the aid of the old capitalist organization. Lenin defended this proposition with the utmost decision
in his April speech on the tasks of the Soviet power. In the

mature circles of the party the idea is not contested. But the idea that we could create an instrument for the defense of the republic, an army, with the aid of the Czarist officers—encountered obstinate resistance. Who could think of re-arming the White officers who had just been disarmed? Thus many comrades questioned. I remember a discussion on this question among the editors of the *Communist,* the organ of the so-called left communists, in which the question of the employment of staff officers nearly led to a split. And the editors of this paper were among the best schooled theoreticians and practicians of the party. It suffices to mention the names of Bukharin, Ossonski, Lomov, W. Yakovlev. There was even greater distrust among the broad circles of our military comrades, recruited for our military organizations during the war. The mistrust of our military functionaries could only be allayed, their agreement to the utilization of the knowledge possessed by the old officers could only be won, by the burning faith of Trotsky in our social force, the belief that we could obtain from the war experts the benefit of their science, without permitting them to force their politics upon us; the belief that the revolutionary watchfulness of the progressive workers would enable them to overcome any counter-revolutionary attempts made by the staff officers.

Trotsky's Magnetic Energy

In order to emerge victorious, it was necessary for the army to be headed by a man of iron will, and for this man to possess not only the full confidence of the party, but the ability of subjugating with his iron will the enemy who is forced to serve us. But Comrade Trotsky has not only succeeded in subordinating to his energy even the highest staff officers. He attained more: he succeeded in winning the confidence of the best elements among the war experts, and in converting them from enemies of Soviet Russia to its most profoundly convinced followers. I witnessed one such victory of Trotsky's at the time of the Brest-Litovsk negotiations. The officers who had accompanied us to Brest-Litovsk maintained a more than reserved attitude towards us. They fulfilled their role as experts with the utmost condescension, in the opinion that they were attending a comedy which merely served to cover a business transaction long since arranged between the Bolsheviki and the German government. But the manner in which Trotsky conducted the struggle against German imperialism, in the name of the principles of the Russian revolution, forced every human being present in the assembly room to feel the moral and spiritual victory of this eminent representative of the Russian proletariat. The mistrust of the war experts towards us vanished in proportion to the development of the great Brest-Litovsk drama.

How clearly I recollect the night when Admiral Altvater— who has since died—one of the leading officers of the old regime, who began to help Soviet Russia not from motives of fear but of conscience, entered my room and said: "I came here because you forced me to do so. I did not believe you; but now I shall help you, and do my work as never before, in the profound conviction that I am serving the fatherland." It is one of Trotsky's greatest victories that he has been able to impart the conviction that the Soviet government really fights for the welfare of the Russian people, even to such people who have come over to us from hostile camps on compulsion only. It goes without saying that this great victory on the inner front, this moral victory over the enemy, has been the result not only of Trotsky's iron energy which won for him universal respect; not only the result of the deep moral force, the high degree of authority even in military spheres, which this social-ist writer and people's tribune, who was placed by the will of the revolution at the head of the army, has been able to win for himself; this victory has also required the self-denial of tens of thousands of our comrades in the army, an iron disci-pline in our own ranks, a consistent striving towards our aims; it has also required the miracle that those masses of human beings who only yesterday fled from the battle-field, take up arms again today, under much more difficult conditions, for the defense of the country.

That these politico-psychological mass factors played an important role is an undeniable fact, but the strongest, most concentrated, and striking expression of this influence is to be found in the personality of Trotsky. Here the Russian revolu-tion has acted through the brain, the nervous system, and the heart of its greatest representative. When our first armed trial began, with Czecho-Slovakia, the party, and with it its leader Trotsky, showed how the principle of the political campaign— as already taught by Lassalle—could be applied to war, to the fight with "steel arguments." We concentrated all material and moral forces on the war. The whole party had grasped the necessity of this. But this necessity also finds its highest expression in the steel figure of Trotsky. After our victory over Denikin in March 1920, Trotsky said, at the party conference: "We have ravaged the whole of Russia in order to conquer the Whites." In these words we again find the unparalleled concentration of will required to ensure the victory. We needed a man who was the embodiment of the war-cry, a man who became the tocsin sounding the alarm, the will demanding from one and all an unqualified subordination to the great bloody necessity.

L. D. Personified the Revolution

It was only a man who works like Trotsky, a man who spares himself as little as Trotsky, who can speak to the soldiers as only Trotsky can—it was only such a man who could be the standard bearer of the armed working people. He has been everything in one person. He has thought out the strategic advice given by the experts and has combined it with a correct estimate of the proportions of social forces; he knew how to unite in one movement the impulses of fourteen fronts, of the ten thousand communists who informed headquarters as to what the real army is and how it is possible to operate with it; he understood how to combine all this in one strategic plan and one scheme of organization. And in all this splendid work he understood better than anyone else how to apply the knowledge of the significance of the moral factor in war.

This combination of strategist and military organizer with the politician is best characterized by the fact that during the whole of this hard work, Trotsky appreciated the importance of Demian Bedny (communist writer), or of the artist Moor (who draws most of the political caricatures for the communist papers, posters, etc.) for the war. Our army was an army of peasants, and the dictatorship of the proletariat with regard to the army, that is, the leading of this peasants' army by workers and by representatives of the working class, was realized in the personality of Trotsky and in the comrades cooperating with him. Trotsky was able, with the aid of the whole apparatus of our party, to impart to the peasants' army, exhausted by the war, the profoundest conviction that it was fighting in its own interests.

Inseparably Linked in History

Trotsky worked with the whole party in the work of forming the Red Army. He would not have fulfilled his task without the party. But without him the creation of the Red Army and its victories, would have demanded infinitely greater sacrifices. Our party will go down in history as the first proletarian party which succeeded in creating a great army, and this bright page in the history of the Russian revolution will always be bound up with the name of *Leon Davidovitch Trotsky*, with the name of a man whose work and deeds will claim not only the love, but also the scientific study of the young generation of workers preparing to conquer the whole world.

1. "UNIFIED

MILITARY DOCTRINE"

November 1, 1921

PREFATORY REMARKS

Comrades, we are now engaged in taking a balance sheet, sifting our ranks and making necessary preparations. Our work in the army has now become minute, detailed, mosaic in character. But it would be unworthy of a revolutionary army to fail to see the forest for the trees. Just because all our efforts in the field of military work are now being directed toward concretizing it, and rendering it more detailed; and because we are turning our attention to partial questions, which make up the whole, we must precisely for this reason tear ourselves away time and again from this detailed work in order to take a survey of the structure of the Red Army as a unity. Here we confront the question of military doctrine and the question of unified military doctrine which are sometimes identified. The conception of military doctrine does not at present appear in a clearly delineated form, nor is it filled with any exact and specific scientific content. The conception of unified military doctrine has been given in part and on the whole a mystical and metaphysical content by those who view it as something akin to an emanation of the national spirit.

Owing to a sharp turn of history a rather natural attempt is being made at present on the plane of the revolutionary class struggle to fill the conception of military doctrine with class content. The realization of this attempt still lies ahead. The greatest vigilance must be exercized here lest one permit himself to be lured into a mystical or metaphysical trap, however it may be disguised with revolutionary terminology; for from a class military doctrine one can only get mysticism and metaphysics, whereas what we want is: A concrete, rich, exact historical conception. For this reason we ask ourselves first of all: Is military doctrine an aggregate of military methods or a theory? Or is it an art, an aggregate of certain applied methods that teach one how to fight?

It is imperative to distinguish between science as the objective knowledge of that which is, and art which teaches how to act.

*   *   *

(Following these prefatory remarks by Leon Trotsky, the first report was delivered by Professor Neznamov. After him the floor was taken by Petrovsky, Verkhovsky, Vatsetis, Tukhachevsky, Svechin and several other active and prominent workers of the Scientific Military Society. Trotsky then summarized.—*Ed.*)

*   *   *

TROTSKY'S SUMMARY SPEECH

Before discussing the gist of the question let me remark that Comrades Verkhovsky and Svechin, while seemingly at opposite poles, stand closest to each other. Comrade Verkhovsky is seized by something akin to terror because, as he says, there is so much discord among us, and we are not agreed on anything and in such a situation it is hardly possible to build anything, let alone gain victory. But, after all, we have built something, and we have not waged war so poorly. I am among those least inclined to idealize the Red Army, but when we had to defend ourselves, we were able to deal blows to our enemies, notwithstanding the discord among us. Comrade Verkhovsky, in my opinion, takes a subjective approach: he overlooks the incontestable—the Red Army's foundation, which no one has questioned and which has actually been erected by the working class. The army once possessed its old summits; there were conscientious and honest elements among the old officerdom, but they have been and are being dissolved. Our army has promulgated a new principle and is creating a commanding staff of a new social origin—a commanding staff, which is perhaps a little bandylegged, insufficiently lit-

erate, but nevertheless endowed with a great historical will.
All of us are guilty of mistakes in theory, but how is it possible not to see the essence of the matter, the *foundation* which
is unconquerable, but to which no one has pointed? What is
there for Comrade Verkhovsky to be afraid of? With his
excellent military virtues, he has nothing to fear.

Groundless Fears

Comrade Svechin says that should some doctrine be invented, he, Svechin, will be made to suffer thereby, because a
censorship will be clamped down. Comrade Svechin, an old
military man who very much reveres Suvorov and Suvorovist
traditions, shies away in fear of censorship. He fears lest
military doctrine prevent the free development of ideas—which
is in part the same thought as Comrade Verkhovsky expressed.
If unified doctrine is understood to mean that there is a ruling
class which has gathered the reins of the army into its own
hands, then none has raised his voice in protest against this.
Let us recall what was written in 1917 and 1918, in our theses,
in our reports to the Soviet Congresses: Their basic idea was to
apply to the country's armed forces the consciousness and the
will of the working class which had founded a new power and
a new state. This is a firmly established fact no longer chal-
lenged even by those who used to dispute it; while those who
tried to fight against it with arms in hand have suffered reverses
and have stopped trying.

For example, there is the volume *Smena Vekh**. These

---

*Smena Vekh—the name of a publication of a tendency among
Russian White Guard emigres, primarily, among their intellect-
ual circles. An anthology, *Smena Vekh*, was published in Prague
in 1921; and presently a periodical of this same name began to
appear there. Both these publications were devoted to an explanation of the motives which had caused the *Smenovekhovtsy*
(literally, "changers of signposts") to pass from a position of
irreconcilable hostility toward the Bolshevik power to one calling
for joint collaboration. These White Guard intellectuals viewed
the New Economic Policy of 1921 (the NEP) as a retreat from
"communism to capitalism"; and argued that this "evolution of
the Bolsheviks" must logically lead to the "reestablishment of
bourgeois relations", and to the strengthening of Russia as a state
capable of defending her independence and interests. They later
issued a daily paper *Nakanune (On the Eve)* in Berlin. This
tendency headed by Ustryalov, Kluchnikov, Potekhin and others
is the true historical originator and precursor of similar theoretical positions since propounded as "original discoveries" by contemporary renegades from Marxism.—*Ed.*

people who once supplied Kolchak with his Ministers have understood that the Red Army is not an invention of a handful of emigres; not a robber's band but the national expression of the Russian people in their present phase of development. And they are absolutely right. None will try deny that a new commanding staff has appeared which is realizing in life the strivings of the toilers, even though in building the army it commits sins against Russian and military literacy. Our misfortune is that the country is illiterate and it will of course require years and years before illiteracy disappears and the Russian toiler begins to commune with culture.

An attempt was made here particularly in Comrade Vatsetis' very rich and valuable speech to give the broadest possible conception of doctrine. Military doctrine embraces everything indispensable for war. War demands that a soldier be healthy; to keep the soldier healthy, in addition to his rations and equipment, a certain hygiene is needed, medicine is required. Herein lies the gist of the thought's aberration. If Clausewitz* said that war is a continuation of politics by other means, then some military men turn this idea around and say that politics is an auxiliary means of war; that all branches of human knowledge are auxiliary sources of military knowledge; and they equate military doctrine with all human knowledge in general. This is absolutely wrong.

The Will To Victory

We are next told that it is necessary to have the desire to fight; it is necessary to have the will to victory. But haven't we all seen the Russian people show this will to victory; haven't we seen it spring to life among the peasants of the Don and the Kuban who have produced their own Budenny, their own cavalry, something quite different from the past when the old Russian nobility used to impose their will on the people? This will to victory has been born even in Russian moujiks, oppressed for ages, let alone the workers. But one must have the will to victory, one must have the desire to fight

---

*Karl von Clausewitz (1780-1831). Prussian general and military theoretician. His best known work "Ueber Krieg und Kriegfuehrung," three volumes Berlin 1832-34, bears unmistakable signs of the influence of the Hegelian dialectic. Clausewitz participated in the campaigns against Napoleon and later served as head of the Prussian General Staff (1831). From 1812-13 he was in the service of the Russian army.—*Ed.*

not for the mere sake of fighting. A great historic goal is indispensable. Czarism had its own goal, and under the previously existing conditions a section of the people espoused it and to some extent developed a will to victory. Well, is there a historic goal in the war at present? Is there such a goal or not? How can any one doubt that there is such a goal, that the present government disposes of detachments of advanced workers who draw the peasantry behind them. It is no accident that we are scoring victories. Therefore there must have been the will to victory. It springs not from military doctrine but from a specific historical task which constitutes the meaning of our entire epoch.

We are also told that it is necessary to know when and why to fight. It is necessary to find one's orientation in the international situation. Well, didn't we find it? Comrade Svechin has said here that the revolutionary epoch is an epoch of empiricism. What shall I say? Never before, in no other country has there been a power so highly theoretical as ours. When still a group of underground emigres we said that the capitalist war would inevitably culminate in revolution. Prior to the revolution we predicted it in theory. What is this if not a theoretical prognosis? The application of science in this field cannot of course be so exact as in astronomy; our calculations are off by perhaps 5 to 10 years. We had hoped for a continuation of the revolution in the West. This did not happen, but nevertheless we did forecast the character of the development.

What does the ill-starred Brest-Litovsk Peace represent? It, too. was an orientation and a theoretical calculation. Our foes had calculated that their own existence was an immutable fact, whereas our existence represented a piece of irrationality; but we held the standpoint of theoretical prognosis and calculated that their days were numbered, whereas our existence remains an immutable fact. I cannot be a military doctrinaire if only for the lack of the necessary military rating, but I did participate together with other comrades in elaborating the following prognosis: It is impossible to fight the Germans and therefore we must make concessions in order to smash them later on. What is this if not an orientation? The knowledge when to fight was supplied us by the basic tenets of Marxism in their application to a given situation. But the desire to fight and the knowledge when to fight still does not provide everything needed for the *ability* to fight. And here military art or knowledge enters into all its rights.

But why is it necessary to drag in absolutely everything under the sun into military knowledge? There are a few things in this world besides military knowledge; there is communism and the world tasks that the working class sets itself; and there is war as one of the methods employed by the working class.

A Few "Innovations"

At this point I must say that those comrades who spoke here in the name of a new military doctrine have completely failed to convince me. I see in it a most dangerous thing: "We'll crush our enemies beneath a barrage of red caps." This happens to be ancient Russian doctrine.

As a matter of fact, what did some comrades say? They said that our doctrine consists not in commanding but in persuading, convincing and impressing through authoritativeness. A wonderful idea! The best thing would be to give Comrade Lyamin 3,000 Tambov deserters and let him organize a regiment with his method. I would very much like to see it done. But how is it possible to do anything at all by a mere stroke of the pen in the face of differences in cultural levels and in the face of ignorance? Our regime is called a regime of dictatorship; we do not conceal this. But some people say that what we need are not commanders-in-chief but commanders-in-persuasion. That's what Kerensky had.

Authoritativeness is an excellent thing, but not very tangible. If one were to impress solely through authoritativeness then what need have we for the *Cheka* and the Special Department? Finally, if we can impress a Tambov moujik solely through our authoritativeness, then why shouldn't we do the same with regard to the German and French peasants?

Comrade Vatsetis reminded us that truth is mightier than force. That is not so. What is correct is only this, that those oppressors who were ashamed of the brute force they applied always covered it up with hypocrisy. Truth is not superior to force; it cannot withstand the onset of artillery. Against artillery only artillery is effective. If you say that the cultural level of peasants and moujiks must be raised, then you are uttering what is an old truth to us. We are all for it and our state apparatus and, in particular, our military affairs must proceed along this line. But it is naive to think that such a task can be solved on the morrow.

We are told that the doctrine of the Red Army comprises

of partisan actions in the enemy's rear and raids deep behind
the front lines. But the first big raid was made by Mamontov*, while Petlura was the leader of partisan formations.
What does this mean? Just how does the doctrine of the Red
Army happen to coincide with the doctrines of a Mamontov
and a Petlura?

Hasty Generalizations

Some comrades have tried to reduce the doctrine of the
Red Army to the use of hand-carts for transport. Inasmuch as
we lack macadam roads and armored trucks, we shall of course
use hand-carts for transportation, that's better than lugging
a machinegun on one's back. But what has military doctrine
to do with it? This is an absolutely incredible manner of pos-
ing the question. Our backwardness and lack of technical
preparation can nowise provide material for military doctrine.

As touches maneuvering, let me point out that we are not
the inventors of the maneuverist principle. Our enemies also
made extensive use of it, owing to the fact that relatively small
numbers of troops were deployed over enormous distances and
because of the wretched means of communication. Much has
been said here about the seizure of cities, points, and so on.
Mamontov captured them from us, and we from him. This
is in the very nature of civil warfare. In one and the same
theater of war, we had our allies behind Mamontov, while in
our midst were Mamontov's allies. Mamontov executed our
agentry; we, his. An attempt is now made to build a doctrine
on this. It is absurd.

Comrade Tukhachevsky sins in the sphere of overhasty generalizations. In his opinion positional warfare is defunct.
This is absolutely wrong. Should we continue to live in peace
conditions for 5 or 10 years—which is not at all excluded—
a new generation will have grown up; the nerve-wracking war
moods under which we labor will have disappeared. A re-
tardation of the revolution in the West would mean a breathing spell for the bourgeoisie. Technology is being restored by

---

*Mamontov was a colonel in Czar's army who became a caval-
ry general in Denikin's White Guard Army. In 1919 Mamontov
gained fleeting fame by the capture of Tambov and his raid into
the Red Army's rear during which his cavalry did great damage,
destroying supplies, supply trains, lines of communication, etc.
—*Ed.*

them as well as by us. We shall be enabled to move up larger and better equipped masses of troops; and with an army of greater mass and better armament there is produced a denser and more stabilized front. An explanation for our excessive maneuvering—which resulted time and again in our advancing 100 versts only in order to retreat 150 versts—is to be found in the fact that the army was so very thin and weak in relation to the given spaces; the armament was so inadequate that the outcome of battles was decided by factors of secondary nature. Why should we seek to hold on to this? What we need is to go beyond this stage of maneuvering which is only the obverse side of guerrilla warfare. I have often recalled that in the first period of the building of our army some comrades said that large formations were no longer needed; that the best thing for us would be a regiment of two or three battalions with artillery and cavalry—and this would comprise an independent unit. Expressed herein was the idea of primitive maneuvering. We have gone beyond this and any idealization of maneuvering would be dangerous in the extreme.

Defense and Offense

It was pointed out here that we must solve problems involving the role of artillery in relation to infantry. In the Kiev area I happened to be present during a heated dispute over the reciprocal relations between artillery and infantry. Every army has hundreds of such questions. This means that on the basis of our civil war experience we must carefully study our statutes and adapt the most important regulations to comply with field conditions. Our statutes must be submitted to a review. We must work them over in our consciousness in terms of our practical experience.

We are proffered a solution to the problem of offense and defense. We are told that our army must take the offensive. There is a great deal of confusion here, and I am afraid that Comrade Tukhachevsky supports in this connection those who are muddling and who say that our army must be an offensive army. Why? Since war is the continuation of politics by other means, therefore our politics should be offensive. But are they? What about Brest Litovsk? And what about our yesterday's declaration of readiness to recognize pre-war debts? It is a maneuver.

Only a daredevil cavalry man is of the opinion that one must always attack. Only a simpleton is of the opinion that

a retreat is tantamount to doom.   Attack and retreat can be
integral parts of a maneuver, and may equally lead to victory.
At the Third World Congress of the Communist International
there was a whole tendency which insisted that a revolutionary
epoch permits only of attack.   This is greatest heresy.   It is
the most criminal heresy which has cost the German proletariat needless blood and didn't bring victory.   Were this tactic
to be followed in the future it would lead to the destruction
of the German revolutionary movement.   In a civil war it is
necessary to maneuver.   And since war is the continuation of
politics by other means, how can we possibly say that military
doctrine always demands the attack?   The Parisian newspaper
Journal des Debats contains an article by a French general
who writes the following:

"In Lorraine we French did the attacking.   As a result of
our offensive the Germans retreated.   But they made a calculated retreat.   They withdrew their front-line forces, moved
up well camouflaged machinegun and artillery posts, and proceeded to annihilate an enormous number of our living forces.
It was a catastrophe.   How did our victory in June 1918 begin?   The German offensive might have proved decisive.   But
we had learned from them in 1914 and employed an elastic
defense, passing over to a counter-offensive after the Germans
had exhausted their forces.   And we smashed the German
army."

You cite the Great French Revolution and its army.   But
don't forget that the French were at the time the most cultured
people of Europe—not only the most revolutionary but the
most cultured and, in point of technology, the most powerful,
provided, of course, we discount England which was powerless
to act on land.   France could permit herself the luxury of offensive politics.   But she crashed none the less.   Although
France did long march triumphantly across Europe, it all
terminated in Waterloo and the restoration of the Bourbons.
But we are among the most uncultured, the most backward
peoples of Europe.   Historical fate compelled us to accomplish
the proletarian revolution in an encirclement of other peoples
not yet seized by it.   Wars lie ahead of us and we must teach
our General Staff to appraise the situation correctly.   Should
we attack or retreat?   Precisely here, knowledge of the most
flexible and elastic kind is required; and it would be the
most colossal blunder for us to impose upon the members
of our General Staff the doctrine: Attack always!   It is the
strategy of adventurism and not revolutionary strategy.

I am likewise in disagreement with the second proposition

advanced by Comrade Tukhachevsky. He considers that the transition to a militia army is incorrect. There are many difficulties in effecting the transition but we are nevertheless passing over to militia forms. In our country with a population of over 100,000,000 we are maintaining an army of one million. This is an approach to a militia. France has 700,000 soldiers, while we have about 1,000,000. Another step in this direction and we shall arrive at a pure militia. We will proceed cautiously because there are difficulties in the reciprocal relations between workers and peasants. But our new policy brings us closer to the peasant and not further away from him. Go to any village you choose, talk there with a moujik and he'll tell you that his attitude toward the Soviet power is friendlier today than it was yesterday. If we grow richer a year hence, and we shall of course grow a little richer, and in two years still richer, this spiral will begin to expand. But even then we shall not act upon the moujik by way of wholesale persuasion as certain young members of our General Staff presume.

In any case, not only persuasions and embraces will have to be employed but also compulsion, although to a lesser degree than hitherto. At the same time much more favorable conditions for organizing a militia will arise among the peasants and the working class. For this reason, doctrine calls only for a reduction of the element of compulsion to lesser proportions than those required in an army of a barracks type. But if we derive our doctrine from the principle that a militia is unnecessary and that what we need is a barracks army, then we shall arrive at all sorts of false metaphysical propositions.

And now, Comrades, I sum up briefly. He speaks the truth who says with regard to the will to victory that the ability is not always to be observed among our commanding staff to develop partial victories and partial successes to full victory. An explanation for this is to be found in the worker-peasant composition of our new commanding staff which inclines to be very easily satisfied with the very first successes attained. But our dispute is over the will to victory in general. I must cite the following example: As all communists know, Turkestan was cut off from the rest of the world, surrounded by Dutovists and other White Guards, but was nevertheless able to hold out for one and a half years without any aid from the outside. What is this if not a manifestation of colossal will to victory? You cannot supply a better example as ground for doctrine. What doctrine other than Marxism can enable one to orient himself in a situation? You should get and read Chicherin's notes and the articles in *Pravda* and *Izvestia*. They point out a correct orientation in the international situation. Take the English *Times* or the French *Le Temps:* Their language is far more exquisite than ours but we orient ourselves in the international situation 100 times better. That's why we have been able to hold out for four years under conditions of blockade and shall continue to hold out.

The Need to Study

Our doctrine is called Marxism. Why invent it a second time? Besides, in order to be able to invent anything except a hand-cart, it is necessary to go to school to the bourgeoisie, once the ability to orient ourselves and the will to victory are given. It is necessary to instill in the minds of our platoon, battalion and division commanders that they must possess not only the will to victory but must also know how to make reports and understand the meaning of maintaining communications, setting up guards, gathering intelligence. And for this the experience of old practice must be utilized. We must study our ABC's. Of no earthly use to us is a military doctrine that declares: "We'll crush our enemies beneath a barrage of red caps." We must eradicate such bravado and revolutionary snobbery. Chaos results whenever strategy is developed from the standpoint of revolutionary youth. Why? Because they have not learned the statutes thoroughly. We looked upon the Czarist statutes with disdain, and thanks to this did not teach them. Yet the old statutes prepare the new.

Marxists have always assimilated the old knowledge; they studied Feuerbach, Hegel, the French encyclopedists and materialists, and political economy. Marx devoted himself to the study of higher mathematics after his hair had grown gray. Engels studied military affairs and natural sciences. It will do incalculable harm if we were to inoculate the military youth with the idea that the old doctrine is utterly worthless and that we have entered a new epoch when everything can be viewed superciliously and with the equipment of an ignoramus.

Elementary Details

Among the younger generation there is of course a revulsion to routine. This is inevitable. But our Academy of the General Staff and the Revolutionary Military Council will

do everything in their power to curb this; and they will be
correct in so doing. I do not look upon this discussion as final.
A few things have been taken down stenographically; we shall
read it over and publish some of it; and perhaps we shall have
other gatherings like this.   Meanwhile, let us not tear ourselves away from elementary needs, rations and boots. I think
that a good ration is superior to a poor doctrine; and as
touches boots, I maintain that our military doctrine begins
with this, that we must tell the Red Army soldier: Learn to
grease your boots and oil your rifle. If in addition to our will
to victory and our readiness to self-sacrifice we also learn
to grease boots, then we shall have the best possible military
doctrine.   And for this reason our attention must be turned
to these practical details.

Now a word concerning technique. Our technique is of
course poor but Europe can't attack us today: Her working
class will not permit it. Hence the conclusion: Europe toler-
ates us. She enters into economic relations with us. Concessions are coming along, although at a steep price. Through
its concessions and trade relations European imperialism will
be compelled to develop our industry and with its own hands
arm us technically against itself. There is no escaping this.
Imperialism is destined to do it, must do it. Were I to say
this publicly before an audience of Lloyd George, Briand and
Millerand, they would shy back in alarm but would nevertheless be constrained to do it, for they have no other way
out. They are driven into relations with us by the European
and world crisis and by the pressure of their working class.
Finally, it is done not by governments but by individual cap-
italists who think of their profits first and always. Hence
flows the conclusion: Don't rush ahead: Comrade Svechin
was correct in saying that time works in our favor. Time is
a very important factor in history. Sometimes a word uttered
five minutes too soon means the loss of a campaign; five
minutes too late is likewise no good; the timing must be exact.
We must now gain a little technical and economic fat. Our
economy is in a state of disruption and recovering very slowly.
We shall have further occasion to debate military doctrine,
clarify our conceptions and render them more precise. The
debate will serve only to advantage in the building of the
Red Army. I propose that in honor of the Red Army we join
in an army cheer!

2. MILITARY DOCTRINE
OR PSEUDO-MILITARY
DOCTRINAIRISM

December 5, 1921

1. OUR METHOD OF ORIENTATION

> "In the practical arts the theoretical leaves and blossoms must not be allowed to grow too high, but must be kept close to experience, their proper soil."—Karl von Clausewitz, *On War* (Theory of Strategy).

A quickening of military thought and a heightening of interest in theoretical problems is unquestionably to be observed in the Red Army. For more than three years we fought and built under fire, then we demobilized and distributed the troops in quarters. This process still remains unfinished to this very day, but the Army is already close to a high degree of organizational definitiveness and has acquired a certain stability. Within it is felt a growing and increasingly urgent need of surveying the road already travelled, drafting the balance sheet, drawing the most necessary theoretical and practical conclusions in order to be better shod for the morrow.

And what will tomorrow bring? New eruptions of civil war fed from without? Or an open attack upon us by bourgeois states? Which ones will strike? How should resistance be prepared? All these questions demand an orientation that is international-political, domestic-political and military-political in character. The situation is constantly changing and, in consequence, the orientation likewise changes. It changes not in principle but in practice. Up to now we have successfully coped with the military tasks imposed upon us by the international and domestic position of Soviet Russia. Our orientation proved to be more correct, more farsighted and deeper-going than the orientation of the mightiest imperialist powers who have sought individually and collectively to bring us down, but who burned their fingers in the attempt. Our superiority lies in possessing the irreplaceable scientific method of orientation —Marxism. It is the most powerful and at the same time subtle instrument—to use it is not as easy as shelling peas. One must learn how to operate with it. Our party's past has taught us through long and hard experience just how to apply the methods of Marxism to the most complex combination of factors and forces during the historical epoch of sharpest breaks. We likewise employ the instrument of Marxism in order to define the basis for our military construction.

It is quite otherwise with our enemies. If the advanced bourgeoisie has banished inertia, routinism and superstition from the domain of productive technology, and has sought to build each enterprise on the precise foundations of scientific methods, then in the field of social orientation the bourgeoisie has proved impotent, because of its class position, to rise to the heights of scientific method. Our class enemies are empiricists, that is, they operate from one occasion to the next, guided not by the analysis of historical development, but by practical experience, routinism, rule of the thumb, and instinct.

Assuredly, on the basis of empiricism the English imperialist caste has set an example of wide-flung predatory usurpation, provided us with a model of triumphant farsightedness and class firmness. Not for nothing has it been said of the English imperialists that they do their thinking in terms of centuries and continents. This habit of weighing and appraising practically the most important factors and forces has been acquired by the ruling British clique thanks to the superiority of its position, from its insular vantage point and under the conditions of a relatively gradual and planful accumulation of capitalist power.

English Empiricists

Parliamentarian methods of personal combinations, of bribery, eloquence and deception, and colonial methods of sanguinary oppression and hypocrisy, along with every other form of vileness have entered equally into the rich arsenal of the ruling clique of the world's greatest empire. The experience of the struggle of English reaction against the Great French Revolution has given the greatest subtlety to the methods of British imperialism, endowed it with utmost flexibility, armed it most diversely, and, in consequence, rendered it more secure against historical surprises.

Nevertheless the exceedingly potent class dexterity of the world-ruling English bourgeoisie is proving inadequate—more and more so with each passing year—in the epoch of the present volcanic convulsions of the bourgeois regime. While they continue to tack and veer with great skill, the British empiricists of the period of decline—whose finished expression is Lloyd George—will inescapably break their necks.

German imperialism rose up as the antipode of British imperialism. The feverish development of German capitalism provided the ruling classes of Germany with an opportunity to accumulate a great deal more in the way of material-technical values than in the way of habits of international and military-political orientation. German imperialism appeared on the world arena as an upstart, plunged ahead too far and came crashing into the dust. And yet not so very long ago at Brest-Litovsk the representatives of German imperialism looked upon us as visionaries, accidentally and temporarily thrust to the top . . .

The art of all-sided orientation has been learned by our party step by step, from the first underground circles through the entire subsequent development, with its interminable theoretical discussions, with its practical measures and failures, attacks and retreats, tactical disputes and turns. Russian emigre garrets in London, Paris and Geneva turned out in the final analysis to be watchtowers of great historical importance. Revolutionary impatience learned to discipline itself through the scientific analysis of the historical process. The will to action became conjoined with restraint and firmness. Our party learned to apply the Marxist method by thinking and doing. And this method serves our party in good stead today . . .

If it may be said of the most far-sighted empiricists of English imperialism that they have a key-ring with a considerable variety of keys good for *many* typical historical situations, then we hold in our hands a universal key which does us service in *all* situations. And while the entire supply of keys inherited by Lloyd George, Churchill and the others is obviously no good for opening a way out of the revolutionary epoch, our Marxist key is predestined above all for this purpose. We are not afraid to speak aloud about this, our greatest advantage over our adversaries, for they are impotent to acquire or to counterfeit our Marxist key.

We foresaw the inevitability of the imperialist war as the prologue to the epoch of proletarian revolution. With this as our starting point we then kept following the course of the war, the methods employed in it, the shifts in the groupings of class forces and on the basis of our observations there crystallized much more directly—if one were to employ a pompous style—the "doctrine" of the Soviet system and the Red Army. From the scientific foresight of the further course of events we gained unconquerable confidence that history is working in our favor. And this optimistic confidence has been and remains the foundation of all our activity.

Marxism does not supply ready-made prescriptions, least of all in the sphere of military construction. But here, too, it provided us with the method. For if it is correct that war is a continuation of politics by other means, then it follows that the army, with bayonets held ready, is the continuation and the capstone of the entire social-state structure.

Our approach to military questions proceeds not from any "military doctrine" as a sum-total of dogmatic postulates; we proceed from the Marxist analysis of what the requirements are for the self-defense of the working class that has taken power into its own hands; the working class that must arm itself after having disarmed the bourgeoisie; that must fight to maintain its power; that must lead the peasants against the landlords; that must not permit the *kulak* democracy to arm the peasants against the workers' state; that must create a reliable commanding staff in the Army, etc., etc.

In building the Red Army we utilized Red Guard detachments as well as the old statutes as well as peasant *atamans* and former Czarist generals. This, of course, might be designated as the absence of "unified doctrine" in the sphere of forming the army and its commanding staff. But such an appraisal would be pedantically banal. Assuredly, we did not take a dogmatic "doctrine" as our starting point. We actually created the army from the historical material ready at hand, unifying all this work from the standpoint of a workers' state fighting to preserve, intrench and extend itself. Those who can't get along without the metaphysically compromised word. doctrine, might say that in creating the Red Army, the armed power on a new class foundation, we thereby built a new military doctrine, inasmuch as despite the diversity of practical measures and the multiplicity of ways and means employed in our military construction, there could not be nor was there either empiricism, barren of ideas, or subjective arbitrariness in the entire work which from beginning to end was fused together by the unity of the class revolutionary goal, by the unity of the will directed to this end, by the unity of the Marxist method of orientation.

2. WITH A DOCTRINE OR WITHOUT?

Attempts have been made and frequently repeated to take the actual work of building the Red Army as a premise for the proletarian "military doctrine." As far back as 1917 the absolute maneuverist principle was counterposed to the "imperialist" principle of positional warfare. The organizational form of the army itself was declared to be subordinate to the revolutionary maneuverist strategy. The corps, the division, even the brigade were proclaimed to be units much too ponderous: The heralds of proletarian "military doctrine" proposed to reduce the entire armed strength of the republic to individual combined detachments or regiments. In essence, this was the ideology of partisan warfare, only slicked up a bit. On the extreme "left," partisan warfare was openly defended. A holy war was declared against statutes, against the old statutes because they were the expression of an out-lived military doctrine; against the new—because they resembled the old too much. True enough, even at that time the adherents of the new doctrine not only failed to provide a draft of new statutes but they did not even present a single article submitting our statutes to any kind of serious principled or rational criticism. Our utilization of the old officers, all the more so their appointment to commanding posts, was proclaimed to be incompatible with the application of the revolutionary military doctrine. And so on and so forth.

As a matter of fact, the noisy innovators were themselves wholly captives of the old military doctrine: The only difference was that they sought to put a minus sign wherever previously there was a plus. All their independent thinking came down to just that. However, the actual work of creating the armed forces of the workers' state proceeded along an altogether different path. We tried—especially in the beginning —to make the greatest possible use of the habits, usages, knowledge and means retained from the past; and we were absolutely

unconcerned about whether the new army would differ greatly
from the old in the formally organizational and technical
sense, or on the other hand, how much resemblance it would
bear to the latter. We built the army with the human and
technical material ready at hand, seeking always and everywhere to render secure the domination of the proletarian vanguard in the organization of the army, that is, in the army's
personnel, its leading staff, its consciousness and in its moods.
The institution of commissars is not some kind of dogma derived from Marxism. Neither is it an integral part of the proletarian "military doctrine." Under specific conditions it simply proved to be an indispensable instrument of proletarian
control, proletarian leadership and political education of the
army, and for this reason it acquired an enormous importance
in the life of the armed forces of the Soviet Republic. The old
commanding staff we combined with the new one; and only
in this way were we able to achieve the necessary result: **The
army proved capable of fighting in the service of the working class.** In its aims, in the class composition of its com-
mander-commissar corps, in its spirit and its entire political
morale, the Red Army differs radically from all other armies
in the world and stands hostilely opposed to them. As the Red
Army continues to develop, it has grown and keeps growing
less and less similar to them in the formal-organizational and
technical fields. Mere exertions to say something new in this
field will not suffice.

Doctrinaire Prejudices

The Red Army is the military expression of the proletarian
dictatorship. Those who require a more solemn formula might
say that the Red Army is the military embodiment of the
"doctrine" of proletarian dictatorship; in the first place, because the proletarian dictatorship is rendered secure by the
Red Army; secondly, because the dictatorship of the proletariat
would be impossible without the Red Army.

The misfortune, however, lies in this, that the awakening
of military-theoretical interests engendered in the beginning a
revival of certain doctrinaire prejudices of the first period of
building the Red Army—prejudices which, to be sure, have
been invested with certain new formulations. but nowise improved thereby. Certain perspicacious innovators have suddenly
discovered that *we are living, or rather not living at all but simply
vegetating without military doctrine,* just like the king in An-

derson's fairy tale who used to go naked without knowing it. There are some who say: "It is high time we created the doctrine of the Red Army." Others sing in chorus: "We haven't been able to find the correct road on all practical questions of military construction for the lack of answers up to the present time to such fundamental questions of military doctrine as: What is the Red Army? What are the historical tasks before us? Will the Red Army have to wage defensive or offensive revolutionary wars? And so on and so forth."

From the way things are put, it turns out that we were able to create the Red Army and, furthermore, a victorious Red Army, but, you see, we failed to supply it with a military doctrine. And this Red Army continues to thrive unregenerate. To the point-blank question of what the doctrine of the Red Army should be, we get the following answer: It must comprise the sum total of the elementary principles of building, educating and applying our armed forces. But this is a purely formal answer. The existing Red Army, too, has its own principles of "building, educating and applying." But under discussion is what kind of doctrine *are we lacking?* That is, what are these *new* principles, which must enter into the program of military construction, and just what is their content? And it is precisely here that the most incredible kind of muddling begins. One individual makes the sensational discovery that the Red Army is a *class* army, the army of proletarian dictatorship. Another one adds to this that inasmuch as the Red Army is a revolutionary, internationalist army, it must be an offensive army. A third proposes in behalf of the spirit of the offensive that we pay special attention to cavalry and aviation. And, finally, a fourth proposes that we don't forget to apply Makhno's hand carts. Around the world in a hand cart—there is a doctrine for the Red Army! It must be said, however, that in all these discoveries any kernel of healthy, not new but correct, ideas is absolutely lost in the husk of idle chatter.

3. WHAT IS MILITARY DOCTRINE?

It is useless to seek for general logical definitions because these will hardly in and by themselves bring us out of the difficulty.* We shall do much better if we approach the question

---

*Comrade Frunze writes: "One may offer a following definition of 'unified military doctrine': It is a unified set of teachings, accepted by an army of a given state, which fix the form of constructing the armed forces of that country, and the methods of

historically. According to the traditional point of view, the foundations of military science are eternal and common to all times and all peoples. But in their concrete refraction these eternal truths assume a national character. Hence are derived: the German military doctrine, the French military doctrine, the Russian military doctrine, and so forth and so on. But if we check the inventory of the eternal truths of military science we obtain little from them beyond a few logical axioms and Euclidian postulates. The flanks must be defended; the means of communication and retreat must be secured; the blow must be directed at the opponent's least defended point, etc., etc. In their essence all these truths, in this all-embracing formulation, transcend far beyond the limits of military art. A donkey in pilfering oats from a torn sack (the opponent's least defended point) and at the same time in turning its rump vigilantly away from the side from which danger may threaten, acts on the basis of the eternal principles of military science. Meanwhile, it is unquestionable that this donkey munching oats has never read Clausewitz.

War, the subject of our discussion, is a social and historical phenomenon which arises, develops, changes its forms and must eventually disappear. For this reason alone war cannot have any eternal laws. The subject of war is man who possesses certain stable anatomical and psychical traits from which flow certain usages and habits. Man operates in a specific and relatively stable geographical milieu. Thus in all wars, during all times and among all peoples there have obtained certain common, relatively stable (but by no means absolute) traits. Based on these traits there has developed historically a military art. Its methods and usages undergo change together with the social conditions which determine it (technology, class structure, forms of state power).

The term "national military doctrine" implied a comparatively stable but nevertheless temporary complex (com-

training and directing the troops (militarily) on the basis of those views which predominate in the given state concerning the character of those military tasks which confront this state and on the basis of the methods of solving them which flow from the class essence of the state and the level of development of its productive forces." (*Krasnaya Nov*, No. 2, page 94. Article of M. Frunze entitled, "Unified Military Doctrine and the Red Army.")

This definition, too, can be accepted conditionally. But as Comrade Frunze's entire article testifies the conclusions drawn from the above-cited definition in no way enrich the ideological arsenal of the Red Army. However, we shall deal with this in greater detail further on.—L. T.

bination) of military calculation, methods, usages, habits, slogans, moods—corresponding to the entire structure of a given society and, first and foremost, the character of its ruling class.

For example, what is the military doctrine of England? Into its composition there evidently enters (or used to enter): the recognition of the urgent need of naval hegemony; a negative attitude toward a regular land army and toward military conscription; or, still more precisely, the recognition of England's need to possess a fleet stronger than the combined fleets of any two other countries and, flowing from this, England's being enabled to maintain a small army on a volunteer basis. Combined with this was the maintenance of such an order in Europe as would not allow a single land power to obtain a decisive preponderance on the continent.

It is incontestable that this English "doctrine" used to be the most stable of military doctrines. Its stability and definitive form were determined by the prolonged, planful, uninterrupted growth of Great Britain's power in the absence of events and shocks that would have radically altered the relationship of forces in the world (or in Europe, which used to signify the selfsame thing *in the past*). At the present time, however, this situation has been completely disrupted. England dealt her "doctrine" the biggest blow when during the war she was compelled to build her army on the basis of compulsory military service. On the continent of Europe, the "equilibrium" has been disrupted. Nobody has confidence in the stability of the new relationship of forces. The power of the United States excludes the possibility for any longer maintaining automatically the rule of the British fleet. It is too early now to forecast the outcome of the Washington Conference.

But it is quite self-evident that after the imperialist war Great Britain's "military doctrine" has become inadequate, impotent and utterly worthless. It has not yet been replaced by a new one. And it is very doubtful that there will ever be a new one, for the epoch of military and revolutionary convulsions and of radical regroupment of world forces leaves very narrow limits for military doctrine in the sense in which we have defined it above with respect to England: *A military "doctrine" presupposes a relative stability of the domestic and foreign situation.*

If we turn to the countries on the continent of Europe, even during the previous epoch, then we find that military doctrine assumes a far less definitive and stable character. What comprised—even during the brief interval of time between the

Franco-Prussian war of 1870-71 and the imperialist war of 1914—the content of the military doctrine of France? Recognition that Germany is a traditional irreconcilable foe; the idea of revenge; educating the army and the new generation in the spirit of this idea; cultivating an alliance with Russia and worshipping the military might of Czarism; and, finally, upholding with none too great assurance the Bonapartist military tradition of bold offensive. The protracted era of armed peace (from 1871 to 1914) nevertheless invested the military-political orientation of France with a relative stability. But the French doctrine was very meager with regard to purely military elements. The war subjected the doctrine of offensive to a cruel test. After the very first few weeks the French army dug into the earth; but although the genuinely French generals and the genuinely French press did not stop reiterating throughout the first period of the war that subterranean trench warfare was a cheap German invention, not at all in harmony with the heroic spirit of the French warrior, the entire war unfolded as a positional struggle of attrition. At the present time the doctrine of the pure offensive, although it has been incorporated into the new French statutes, is being, as we shall see, sharply opposed in France herself.

The military doctrine of post-Bismarck Germany was much more aggressive in its essence, in correspondence with the country's politics, but much more cautious in its strategic formulations. "Principles of strategy never transcend common sense," was one of the mottoes of the German highest military school for commanders. However, the rapid growth of capitalist wealth and population lifted the ruling circles of Germany, particularly her noble-officer caste to ever greater heights. Germany's ruling classes lacked the experience of operating on a world scale; they failed to take forces and resources properly into account, and invested their diplomacy and strategy with a super-aggressive character, far removed from "common sense." German militarism fell victim to its own unbridled aggressiveness.

What follows from this? It follows that the term, national doctrine, implied during the last epoch a complex of stable, guiding diplomatic and military-political ideas and of strategical directives more or less bound up with the former. In addition, the so-called military doctrine—the formula for the military orientation of a ruling class of a given country in international conditions—attains greater definitiveness of form the more definitive, stable and planful is the domestic and international position of the country throughout its development.

The imperialist war and the resulting epoch of greatest instability have in all fields absolutely cut the ground from under national-military doctrines, and have placed on the order of the day the necessity of swiftly taking the changing situation into account, with its new groupings and new combinations, with its "unprincipled" tacking and veering under the sign of the current troubles and alarms. In this connection the Washington Conference provides a very instructive picture. It is absolutely incontestable that today after the test to which the old military doctrines have been submitted by the imperialist war, not a single country has retained principles and ideas stable enough to be designated as a national military doctrine.

One might, it is true, venture to presuppose that national military doctrines will once again take shape as soon as a new world relationship of forces is established, together with the corresponding position of each particular state. This presupposes, however, that the revolutionary epoch of shocks and convulsions will be liquidated and then replaced by an epoch of organic development. But there is no ground whatever for such a supposition.

4. COMMONPLACES AND IDLE CHATTER

It might seem that the struggle against Soviet Russia ought to be rather a stable element of the "military doctrine" of all capitalist states in the present epoch. But even this is not the case. The complexity of the world situation, the monstrous crisscrossing of contradictory interests and, primarily, the unstable social foundations of bourgeois governments exclude the possibility of consistently carrying out even a single military "doctrine"—the struggle against Soviet Russia. Or, to put it more precisely, the struggle against Soviet Russia changes its form so frequently and unfolds along such zigzags that the mortal danger for us lies in lulling our vigilance with petty doctrinaire words and "formulas" involving international relations. The sole correct "doctrine" for us is: *Be on guard and keep both eyes open!* It is impossible to give an unconditional answer even when the question is posed in its crudest form, namely: Will our chief arena of military activity in the next few years be in the West or in the East? The world situation is far too complex. The general course of historical development is quite clear, but events do not follow an order fixed in advance, neither do they mature according to a set schedule. In practice one must react not to the "course of development" but to facts, to events. It is not difficult to conjure up historical variants which would compel us to engage our forces primarily in the East, or, conversely, in the West, coming to the aid of revolutions; conducting a defensive war, or on the other hand, finding ourselves compelled to pass over to the offensive. Only the Marxist method of international orientation, of calculating the class forces in all their combinations and shifts can enable us to find a proper solution in each given concrete case. It is impossible to invent a general formula that would express the "essence" of our military tasks in the next period.

One can, however—and this is not infrequently done—endow the concept of military doctrine with a far more concrete and narrow content, by restricting its meaning to those elementary principles of purely military affairs which regulate all the aspects of military organization, tactics and strategy. In this sense it may be said that the content of military statutes is determined directly by military doctrine. But what kind of principles are these? Certain doctrinaires depict the matter as follows: It is first necessary to establish the essence and purpose of the army and the task before it; from this definition one then derives the army's organization, its strategy and tactics; and incorporates these deductions in statutes. In reality, such an approach to the question is scholastic and lifeless.

An inkling of the assortment of banalities and idle chatter that are subsumed under the elementary principles of military art may be gleaned from the solemnly quoted statement of Foch to the effect that the essence of modern war consists in "once the hostile armies are located in destroying them, employing to this end the direction and tactics which lead most quickly and surely to the desired goal." How profound! What boundless horizons this opens before us! To amplify this one need only add that the essence of modern methods of nutrition consists in locating the aperture of the mouth, introducing food therein, and, after it has been masticated with the least possible expenditure of energy—in swallowing it. Why shouldn't one try to deduce from this principle—which is in no way inferior to that of Foch—precisely what the food is and how it must be prepared and just when and just who should swallow it; and, above all, how this food is to be procured.

Military affairs are very empirical, very practical affairs. It is a very risky exercise to attempt to erect them into a system from whose fundamental principles are to be deduced field statutes and the structure of squadrons and the cut of the uniform. This was very well understood by old Clausewitz who said:

> "It is not impossible, perhaps, to write a systematic theory of war, both logical and wide in scope. But our theory, up to the present, is far from being either. Not to mention their unscientific spirit in the attempt to make their systems consistent and complete, many such works are stuffed with commonplaces and idle chatter of every kind."

5. HAVE WE A "MILITARY DOCTRINE"?

Do we need a "military doctrine" or don't we? I have been accused by some of "evading" an answer to this question. But after all in order to give an answer one must know what is being asked, that is, just what is meant by military doctrine. So long as the question is not posed clearly and thoroughly, one cannot help "evading" an answer. In order to come closer to the correct formulation of the question let us, after everything that has been said, separate the question itself into its component parts. From this point of view, "military doctrine" may be said to consist of the following elements:

1. The fundamental (class) orientation of our country followed by the government in the questions of economy, culture, etc., that is, in domestic policy.

2. The international orientation of the workers' state. The most important lines of our world policy and, tied up with the latter, the possible theatres of our military activities.

3. The personnel and construction of the Red Army in correspondence with the nature of the workers'-peasants' state and the task of its armed forces.

4. The strategical and tactical schooling of the Red Army.

The tenets relating to the organization of the army (point 3) together with those relating to the strategic schooling (point 4) must constitute, as is self-evident, the military doctrine in the proper (or narrow) sense of the term.

One could proceed to subdivide still further. For example it is possible to separate out from the enumerated points the question pertaining to the technology of the Red Army, or the manner in which propaganda work is carried on, and so forth and so on.

Must the government, the leading party, the military department have definite views on all the questions? Why, of course they must. How is it possible to build the Red Army without having definite views on its social composition, on the personnel of the officer-commissar corps, on the training and education of the various branches, etc.? Conversely, it is impossible to obtain answers to these questions without probing into the fundamental domestic and international tasks of the workers' state. In other words the military department must be equipped with certain guiding principles on which the army is built, trained and reorganized.

Is it necessary (and should one) designate the sum-total of these principles as military doctrine?

To this my answer has been and remains: If anyone wants to label the sum-total of the principles and the practical methods of the Red Army as military doctrine, then, although I do not share this passion for the withered adornments of ancient officialdom, I will not fight over it (my inclination is to evade such fights). But if anyone is so bold as to assert that we do not possess these elementary principles and practical methods*, and that we have not worked and are not working collectively on this, then my answer is: You are not speaking the truth; you are befuddling yourselves and others with idle chatter. Instead of screaming about military doctrine you should present us with it, demonstrate it, show us at least a particle of that military doctrine which the Red Army is presumably lacking. But the whole trouble is that as soon as our military "doctrinaires" pass from lamentations about the usefulness of doctrine to actual attempts to present us with one, or even with its most general outline they either repeat, not very adequately, what has long ago been said, what has already been assimilated by us, what has already been incorporated in resolutions of the party and of the Soviet congresses, in decrees, regulations, statutes and instructions, far better and much more precisely than is done by our alleged innovators; or they confuse things, commit blunders and indulge in absolutely impermissible "independent thinking."

We shall now proceed to prove this with regard to each one of the component elements of the so-called military doctrine.

---

*Comrade Solomin accuses us (See the military-scientific journal, *Military Science and the Revolution*) of having failed as yet to give an answer to the question:"What kind of army are we preparing and for what tasks?"—L. T.

6. "WHAT KIND OF ARMY ARE WE PREPARING AND FOR WHAT TASKS"?

"The old army served as the instrument for the class oppression of the toilers by the bourgeoisie. With the transfer of power to the toiling and exploited classes, the necessity has arisen of creating a new army which would at present serve as the bulwark of Soviet power and which would in the near future provide the basis for replacing the regular army by the armed people, and give support to the impending socialist revolution in Europe."

So reads the decree on the formation of the Red Army issued by the Council of Peoples Commissars on January 12, 1918. I am very sorry that it is impossible to adduce here everything that was said concerning the Red Army in our party program and in the resolutions of our congresses. Let me urge the reader to reread them. They are both useful and instructive. In them it is very clearly stated "what kind of army we are preparing and for what tasks." What are the newly-baked military doctrinaires preparing to add in this connection? Instead of wracking their brains in order to rehash precise and clear formulations, they would do much better to devote themselves to clarifying these formulas through propaganda work among young Red Army soldiers. This work is far more fruitful.

But it may be said—and it is said—that the resolutions and decrees do not sufficiently underscore the *international* role of the Red Army, and, especially, the need of preparing for offensive revolutionary wars. Solomin is very emphatic on this point... On page 22 of his article, Solomin writes:

"We are preparing the class army of the proletariat, a worker-peasant army, not only for the defense against the bourgeois-landlord counter-revolution but also for revolutionary wars (both defensive and offensive) against the imperialist powers, and for wars of semi-civil(?) type in which offensive strategy can play a major role."

Such is the revelation, almost the revolutionary gospel of Solomin! But,—alas!—as is often the case with apostles, our author is cruelly mistaken in thinking that he has discovered something new. He is only formulating poorly something quite old. Precisely because war is a continuation of politics with rifles in hand there was not and could not be in our party any principled controversy over the place which revolutionary wars can and must occupy in the development of the world prole-

tarian revolution. This question has been posed and solved in
the Russian Marxist press quite a while ago. I could adduce
dozens of leading articles from the party press, especially following the outbreak of the imperialist war, which treat of the
revolutionary war of the workers state as something to be taken
for granted. But I will go back even further and cite lines
which I had the occasion to write in 1905-06.

"From the very beginning, this (development of the Russian
revolution) will invest the unfolding events with an international character and will open up the most grandiose per-
spective: The political emancipation under the leadership of
the working class of Russia will raise this leading class to
unheard of heights in history, transfer into its hands colossal
forces and resources, and make it the initiator of the world
liquidation of capitalism for which history has already provided
all the necessary objective premises.

"Should the Russian proletariat, temporarily taking power
into its hands, fail of its own initiative to spread the revolution
to the soil of Europe, it will be *compelled* to do so by the
European feudal-bourgeois reaction.

"Of course, it would be an idle thing to speculate at the
present time about the paths through which the Russian revolution will be transmitted to old capitalist Europe: These
paths may prove to be completely unexpected. *For the sake of
illustrating our thought rather than as a forecast let us take
Poland as the connecting link between the revolutionary East
and the revolutionary West.*

"The triumph of the revolution in Russia would inevitably
signify the victory of the revolution in Poland. It is not difficult
to imagine that the establishment of the revolutionary regime
in the nine Polish provinces held by Russia will inevitably raise
Galicia and Poznan to their feet.* The governments of the
Hohenzollerns and of the Habsburgs will reply to this by de-
ploying their military forces at the Polish border in order then
to cross it and to crush the enemy in the center—at Warsaw.
Clearly, the Russian revolution will not be able to leave its
Western vanguard in the hands of the Prusso-Austrian troops.
In these conditions, war against the governments of Wilhelm II
and Franz-Joseph will be dictated to the revolutionary government of Russia by the law of self-preservation. What position
will the German and Austrian proletariat then take? Clearly,
they will not be able to remain calm observers of this counter-
revolutionary crusade of their national armies. The war of
feudal-bourgeois Germany against revolutionary Russia will
inevitably signify the proletarian revolution in Germany. To
those to whom such an assertion may seem too categoric we

*Let me recall that this was written in 1905.—L. T.

propose that they try to conceive of another historical event more likely to impel the German workers and the German reaction onto the road of measuring their forces openly." (See, *Our Revolution, by Leon Trotsky, p. 280.*)

Naturally, the events have not unfolded in the historical order indicated tentatively for the purpose of illustration in these lines written sixteen years ago. But the main course of development has confirmed and continues to confirm the prognosis to the effect that the epoch of proletarian revolution must inescapably become the epoch of revolutionary wars; and that the conquest of power by the young Russian proletariat will inevitably propel it into war with the forces of world reaction. Thus, more than a decade and a half ago we already clearly understood in essence "what kind of army, and for what tasks" we had to prepare.

7. REVOLUTIONARY POLITICS AND METHODISM

For us, no *principled* question is involved with regard to offensive revolutionary warfare. But so far as this "doctrine" is concerned, the proletarian state must say what has been said by the last World Congress of the CI concerning the revolutionary offensive (the doctrine of the offensive) of the working masses in bourgeois states: Only a traitor can renounce the offensive; only a simpleton will reduce our entire strategy to the offensive.

Unfortunately, there are not a few *simpletons of the offensive* among our new-fashioned doctrinaires who, under the banner of a military doctrine, are seeking to introduce into our military circulation the same unilateral "leftist" tendencies which at the Third World Congress of the Comintern attained their fruition in the guise of the theory of the offensive: *Inasmuch as* (!) we are living in a revolutionary epoch, *therefore* (!) the Communist Party must carry out the policy of the offensive. To translate "leftism" into the language of military doctrine, is to multiply the error manyfold. While safeguarding the principled ground of waging an irreconcilable class struggle, Marxist tactics are at the same time distinguished by utmost flexibility, mobility, or, to speak in military language, maneuverability. To this principled firmness, flexible in methods and forms, there is counterposed a rigid methodism, which transforms into

an absolute method such questions as our participation or non-
participation in parliamentary work, our acceptance or rejection of agreements with non-communist parties and organizations—an absolute method presumably applicable to any and
all circumstances.

The word "methodism" is most frequently employed in
military-strategic literature. Characteristic of epigones, of
mediocre army leaders and routinists is the attempt to erect
into a stable system a certain combination of actions, corresponding to a specific set of conditions. Inasmuch as war is
not waged by men constantly, but only after considerable inter-
ruptions, it is a common phenomenon to find the methods and
usages of the last war holding sway over the consciousness of
military workers during the periods of peace. That is why
methodism is revealed more graphically in the military sphere.
The false tendencies of methodism unquestionably find their
expression in attempts to construct the doctrine of "offensive
revolutionary war."

This doctrine contains two elements: International-political
and the operative-strategic. For it is a question, in the first
place, of unfolding through the language of war an offensive
international policy for the sake of hastening the revolutionary
culmination; and, secondly, of investing the strategy of the Red
Army itself, with an offensive character. It is necessary to
separate these two questions even though they are mutually
connected in certain relations.

That we do not renounce revolutionary wars is attested not
only by articles and resolutions but by major historical facts.
After the Polish bourgeoisie imposed upon us a defensive war
in the Spring of 1920, we made the attempt to develop our
defense into a revolutionary offensive. True, our attempt was
not crowned with success. But hence flows the not unimportant
supplementary conclusion that revolutionary war, the incontestable instrument of our policy under certain conditions, can
—under other conditions—lead to results opposite to those
intended.

In the Brest-Litovsk period we were constrained for the first
time to apply on a broad scale the policy of political-strategical
retreat. It seemed to many at the time that this would prove
fatal to us. But within a few months it was demonstrated that
time had worked excellently in our favor. In February 1918
German militarism, while already undermined, nevertheless still
remained strong enough at the time to crush us and our insignificant military forces. In November German militarism fell

apart. Our international-political Brest retreat was our salvation.

After Brest we were compelled to wage uninterrupted war against the White Guard armies and the foreign interventionist detachments. This small-scale war was defensive and offensive both politically and militarily. On the whole, however, the foreign policy of our government during that period was primarily the policy of defense and of retreat (no sovietization of the Baltic states, our frequent offers to enter into peace negotiations along with our readiness to make the biggest concessions, the "new" economic policy, recognition of Czarist debts, etc.). In particular, we were most conciliatory in our relations with Poland offering her better conditions than those projected by the Allies. Our efforts were not crowned with success. Pilsudski attacked us. The war clearly assumed a defensive character on our side. This fact aided in the extreme to rally the public opinion not only of workers and peasants but also of many bourgeois-intellectual elements. Successful defense naturally developed into a victorious offensive. But we over-estimated the internal revolutionary potentialities of Poland at that time. This over-estimation found its expression in the excessive aggressiveness of our operations, that is, an aggressiveness beyond our resources. We advanced too far and the result is well known: we were thrown back.

Almost simultaneously with this, the mighty revolutionary wave in Italy was broken not so much by the resistance of the bourgeoisie as by the perfidious passivity of the leading workers' organizations. The failure of our August offensive against Warsaw and the crushing of the September movement in Italy altered the relationship of forces in favor of the bourgeoisie of entire Europe. From that time on there is to be observed a greater stability in the political position of the bourgeoisie and a greater assurance in its conduct. The attempt of the German Communist Party to hasten the revolutionary culmination through an artificial general offensive did not produce and could not have produced the desired results. The revolutionary movement has proceeded at a far slower tempo than we expected in 1918-1919. The social soil, however, remains mined. The commercial-industrial crisis assumes ever more monstrous proportions. Abrupt shifts in the political development in the form of revolutionary explosions, are wholly possible in the immediate future. But, on the whole, the development has become more sluggish. The Third World Congress of the International has summoned the communist parties to make careful and stubborn preparations. In many countries the Communists have been obliged to carry out major strategic retreats, and to renounce the immediate solution of combat tasks they had recently set themselves. The initiative of the offensive has passed temporarily into the hands of the bourgeoisie. The work of the communist parties is now primarily defensive and preparatory-organizational in character. Our revolutionary defense remains as always elastic and firm, that is, capable of becoming transformed, with a corresponding change in conditions, into a counter-offensive which in its turn can lead to the decisive battle.

The failure of the offensive against Warsaw, the victory of the bourgeoisie in Italy, and the temporary ebb in Germany have compelled us to execute a sharp retreat, beginning with the Riga Treaty and terminating in a conditional recognition of Czarist debts.

During this same period we executed a retreat of no lesser proportions in the sphere of economic construction: the authorization of concessions, the abolition of grain monopoly, the leasing of many industrial enterprises, etc. The basic reason for these successive retreats is to be found in the maintenance of the capitalist encirclement, that is, the relative stability of the bourgeois regime.

Just what is it that the proponents of military doctrine want (for the sake of brevity we call them doctrinaires—a designation they have earned), who demand that we orient the Red Army from the standpoint of offensive revolutionary war? Do they simply want the bare recognition of the principle? In that case they are breaking into open doors. Or do they consider that in the international situation or in our domestic situation such conditions have arisen as place an offensive revolutionary war on the agenda? But in that case our doctrinaires should aim their blows not at the military department but at our party and the Communist International, for it was none other than the World Congress of the CI that rejected in the summer of this year the offensive revolutionary strategy as untimely, summoned all parties to undertake careful preparatory work and approved the defensive-maneuverist policy of Soviet Russia, as a policy imposed by the objective conditions.

Or do some of our doctrinaires perhaps consider that while the "weak" communist parties in bourgeois countries must carry on preparatory work, the "all-powerful" Red Army ought to undertake an offensive revolutionary war? Are there perhaps some impatient strategists who really want to transfer onto the

shoulders of the Red Army the burden of the "final and decisive conflict" in the world or in Europe alone? Whoever
seriously propagates such a policy had better hang a millstone
about his neck and proceed in accordance with the subsequent
Biblical instruction.

8. EDUCATION "IN THE SPIRIT" OF THE OFFENSIVE

Seeking to extricate himself from contradictions involved in
a doctrine of the offensive during an era of defensive retreat,
Comrade Solomin invests the "doctrine" of revolutionary war
with—an educational meaning. At the present time, he con-
cedes, we are very much interested in peace and will do everything in order to preserve it. But revolutionary wars, despite
our defensive policy, are inevitable. We must prepare for
them, and consequently must instill—through education—an
offensive "spirit" for future use. The offensive is, therefore,
to be understood by us not in a material but spiritual sense.
In other words, along with a reserve supply of army biscuits,
Comrade Solomin wants to have a reserve supply of offensive
enthusiasm. Things get worse and worse from one hour to the
next. If, as we have seen from the foregoing, our severe critic
lacks an understanding of revolutionary strategy, then he dem-
onstrates here a lack of understanding of the laws of revolutionary psychology.

We need peace not because of doctrinal considerations but
because the toilers have been exhausted by war and privations.
We are striving to safeguard as long a period of peace as possible for the workers and the peasants. We explain to the Army
that if we do not demobilize it is only because new attacks
threaten us. From these conditions Solomin draws the conclusion that the Red Army must be "educated" in the ideology
of offensive revolutionary war. What an idealistic approach to
"education"! "We have not the strength to wage war," Comrade
Solomin reasons mournfully, "nor do we intend to wage war,
but we must be prepared, and therefore we must prepare for
the offensive—such is the contradictory formula which we arrive
at." This formula is indeed contradictory. But Solomin is very
much mistaken if he thinks that this is a "good", a dialectic
contradiction; this is pure and simple muddling.

One of the most important tasks of our domestic policy in
the last period has been to draw closer to the peasant. The
peasant question confronts us with special sharpness in the

Army. Does Solomin seriously believe that today after the
immediate danger of landlordism has been eliminated and while
the European revolution still remains a potentiality, we can
weld together an army of more than one million, nine-tenths
peasants, under the banner of offensive war to bring the proletarian revolution to its culmination? Propaganda of this kind
would be stillborn.

We do not of course intend to hide for a moment from the
toilers, including the Red Army, that we shall always remain in
principle in favor of offensive revolutionary war, under conditions when such a war can aid to emancipate the toilers of other
countries. But to believe that on the basis of this principled
declaration it is possible to create an actual ideology or to
"educate" the Red Army under the existing conditions is to
understand neither the Red Army nor the existing conditions.
As a matter of fact, every sensible Red soldier is convinced
that if we are not attacked during winter or spring, we shall
not, in any case, disturb the peace, but exert all our efforts in
order to heal our wounds, in order to utilize the breathing spell.
In our exhausted country we are learning the military art, arming ourselves, building a large army in order to defend ourselves against attack. Here is a "doctrine"—clear, simple, corresponding to reality.

It was precisely because in the spring of 1920 we posed the
question in this manner that every Red soldier became firmly
convinced that bourgeois Poland imposed upon us a war which
we did not want and against which we tried to safeguard the
people by our readiness to make the greatest concessions. This
conviction gave birth to the greatest indignation and hatred
against the enemy. It was precisely owing to this that the war,
beginning as one of defense, was later able to unfold as an
offensive war.

The contradiction between defensive propaganda and the
offensive character of war—offensive, in the last analysis—is
a "good," viable, dialectic contradiction. We have no grounds
whatever for changing the character and direction of our military educational work in order to please muddleheads, even if
they speak in the name of military doctrine.

Those who talk about revolutionary wars most frequently
gather their inspiration from recollections of the wars of the
Great French Revolution. In France they also began with
defense; created an army on the basis of defense and later
passed to the offensive. To the tune of the *Marseillaise* the
armed sansculottes swept over Europe like a revolutionary
tornado.

Historical analogies are very tempting. But it is necessary to be careful in employing them. Otherwise, misled by the formal traits of resemblance, one may overlook the material traits of difference. At the end of the eighteenth century France was the richest and most civilized country on the European continent. In the twentieth century Russia is one of the poorest and most backward countries in Europe. Compared with the revolutionary tasks which now confront us, the revolutionary task of the French Army was much more superficial in character: At the time it was a question of overthrowing "tyrants"; it was a question of abolishing or mitigating feudal servitude. Nowadays it is a question of completely destroying exploitation and class oppression.

But the role of French arms, that is of an advanced country in relation to backward Europe, proved to be very limited and transitory even in relation to bourgeois-revolutionary tasks. With the downfall of Bonapartism which had grown out of the revolutionary war, Europe returned to its kings and feudal lords.

In the gigantic class struggle unfolding today, the role of military intervention from the outside can acquire only a supplementary, contributory, auxiliary significance. Military intervention can hasten the culmination and facilitate the victory. But this cannot occur unless the revolution is mature not only with regard to social relations—and this condition is already fulfilled—but also with regard to political consciousness. Military intervention may be likened to the forceps of an obstetrician, which if applied in time can reduce the birth pangs, but if brought into play prematurely can produce only a miscarriage.

What we have said up to now applies not so much to the Red Army, its construction and methods of operation as to the political tasks set for the Red Army by the workers' state.

Let us now approach military doctrine in the more narrow sense of the term. We have heard from Comrade Solomin that so long as we fail to adopt the doctrine of offensive revolutionary war, we shall continue to muddle and to commit blunders in organizational, military-pedagogical, strategical and other questions. However, such a commonplace gets us nowhere. Instead of repeating that good practical conclusions must necessarily flow from a good doctrine, the thing to do is to present us with these conclusions. Alas! No sooner do our doctrinaires attempt to reach conclusions than they offer us either a pathetic rehash of elementary truisms or the most pernicious products of "independent thinking."

9. THE STRATEGICAL AND TECHNICAL CONTENT OF "MILITARY DOCTRINE" (MANEUVERABILITY)

Our innovators devote their greatest energies to an attempt to anchor military doctrine in the sphere of operational questions. According to them, strategically the Red Army differs *in principle* from all other armies inasmuch as in our epoch of positional immobility the basic features of the Red Army's operations are: *maneuverability and aggressiveness.*

The operations of the civil war are unquestionably distinguished by extraordinary maneuverability. But here it is first necessary to give the most precise answer to the following question: Does the maneuverability of the Red Army flow from its inner qualities, its class nature, its revolutionary spirit, its fighting zeal or does it, on the contrary, flow from the objective conditions, the vastness of the military theaters and the relatively small number of troops employed? This question is of no small importance, especially if we grant that revolutionary wars will be waged not only on the Don and the Volga but also on the Seine, the Scheldt and the Thames.

But let us meanwhile return to our native rivers. Was the Red Army alone distinguished by maneuverability? No. The strategy of the Whites was without exception maneuverist. In most instances their troops were inferior to ours in numbers and in point of morale, but they were superior in military skill. Hence the need of maneuverist strategy was felt most urgently by the Whites. During the initial stages we learned about maneuverability from them. In the final stage of the civil war we invariably witnessed maneuver against maneuver. Finally, the operations of Ungern's and Makhno's detachments—these degenerate, bandit outgrowths of the civil war—were distinguished by the greatest maneuverability. What conclusion follows from this? It follows that maneuverability is not peculiar to a revolutionary army but to civil war as such.

In national wars, a fear of distances accompanies the operations. By removing itself from its base, from its own people, from the sphere of its own language, an army or a detachment falls into a completely alien environment where neither support, cover nor assistance is available. In a civil war each side finds sympathy and support to a greater or lesser degree in the opponent's rear. National wars are waged (at all events, they used to be waged) by huge masses with all the national-state resources on both sides being brought into play. Civil war signifies that the forces and resources of the country that is convulsed by revolution are divided in two parts; that warfare, especially in the first stage, is waged by an initiatory minority on each side, and consequently by masses of far lesser bulk and greater mobility; and for this reason improvisation and accident play a much more decisive part.

Civil war is characterized by maneuverability in both camps. It is consequently impermissible to consider maneuverability as the peculiar expression of the revolutionary character of the Red Army.

We conquered in the civil war. There are no grounds whatever for doubting that the superiority of the strategic leadership was on our side. In the final analysis, however, victory was assured by the enthusiasm and self-sacrifice of the proletarian vanguard and the support of the peasant masses. But these conditions are not created by the Red Army but represent the historical preconditions for its rise, its development and its successes.

Comrade Varin remarks in the magazine *Military Science and the Revolution* that the mobility of our troops surpasses all historical precedents. This is a very interesting assertion. It ought to be carefully verified. It is unquestionable that the extraordinary speed of operational movements, demanding endurance and self-sacrifice, was conditioned by the revolutionary spirit of the Army, by the zeal the Communists introduced. For the students of our Military Academy it would be a most interesting assignment to compare the marches of the Red Army from the standpoint of distances covered with other historical examples, particularly the campaigns of the armies of the Great French Revolution. On the other hand, a comparison should be made of the very same elements as they relate both to the Reds and Whites in our civil war. When we attacked, they retreated, and vice versa. Did we actually show, on the average, greater endurance during campaigns; and to what extent was this a factor in our victory? It is incontestable that the Communist leaven could produce a superhuman exertion of forces in individual cases. But it would take a special investigation to determine whether the same result would hold for an entire campaign in the course of which the limits of physiological capacity could not fail to manifest themselves. Such an investigation does not of course promise to turn all strategy topsy-turvy. But it would undoubtedly enrich with certain valuable factual data our knowledge of the nature of civil war and of the revolutionary army.

Attempts to fix as laws or to erect into a dogma those features of the Red Army's strategy and tactics which have characterized it in the last period can prove most harmful and even fatal. It is possible to say in advance that the operations of the Red Army on the Asiatic mainland—if they are destined to unfold there—would of necessity be profoundly maneuverist in character. The cavalry would have to play the most important role, and in certain cases, the one and only role. But, on the other hand, there can be no doubt that military activities on the Western theater would be far more restricted in character. Operations conducted on territories with a different national composition and more thickly populated—with greater masses of troops per given area—would undoubtedly bring the war close to a positional one, and in any case would impose far narrower limits on the freedom of maneuver.

The recognition that it is inadmissable for the Red Army to defend fortified positions (Tukhachevsky) sums up correctly, in part and on the whole, the lessons of the last period, but it cannot, in any case, be recognized as an unconditional rule for the future. Defense of fortified positions demands fortress troops, or more correctly, highly trained troops, fused by experience and confident of themselves. In the last period we only began to accumulate this. Every regiment as well as the entire army in general was a living improvisation. It was possible to assure enthusiasm and zeal—and we secured it but it was impossible to create artificially the necessary routine, the automatic fusion of the neighboring sections and their confidence in mutually assisting one another. It is impossible to create traditions by decree. To a large extent this does exist now and we shall accumulate more and more as time goes on. Thereby we obtain the prerequisites both for better carrying out maneuverist operations and, if need arises, positional actions.

We must reject all attempts at building an absolute revolutionary strategy with the elements of our limited experience of three years of civil war during which army sections of a special quality engaged in combat under special conditions. Clausewitz has warned very correctly against this. He wrote:

"What is more natural than that the revolutionary war (of France) had its own way of doing things? And what theory could have included that peculiar method? The trouble is that such a manner, originating from a special case easily outlives its day, because it continues *unchanged* while circumstances imperceptibly undergo complete *change*. That is what theory should prevent by lucid and rational criticism. In 1806 the victims of this methodism were the Prussian generals . . ."

Alas! The Prussian generals are not the only ones who incline toward methodism, i.e., platitudes and stereotypes.

10. OFFENSE AND DEFENSE IN THE LIGHT OF THE IMPERIALIST WAR

It is proclaimed that the second specific trait of revolutionary strategy is *aggressiveness*. The attempt to build a doctrine upon this turns out to be all the more one-sided in view of the fact that during the epoch prior to the first World War the strategy of offense was nurtured in the by no means revolutionary general staffs and military academies of almost all the big countries of Europe. Contrary to what Comrade Frunze writes * the offense was (and formally still remains to this very day) the official doctrine of the French Republic. Jaures tirelessly fought against the doctrinairism of pure offense, counterposing to it the pacifist doctrinairism of pure defense. The sharp reaction against the traditional official doctrine of the French General Staff came as a result of the last war. It might not be amiss to cite here two graphic pieces of evidence. The French military journal, *Revue Militaire Francaise* (September 1, 1921, page 336) adduces the following proposition borrowed from the Germans and incorporated by the French General Staff in 1913 into "The Statute on the Conduct of Combat Actions by Large Units." This proposition reads:

"The lessons of the past have brought their fruits: *The French Army, returning to its traditions, henceforth does not permit of conducting operations under any law other than that of offense.*"

The military journal goes on to comment:

"This law shortly thereafter introduced into our statutes on general tactics and on the partial tactics of different kinds of arms was made the basis of our entire military science which was implanted in the minds both of the students of our General Staff as well as of our commanding corps through joint discussions, practical exercises on maps or in the field and, finally, through the so-called *major maneuvers.*"

"This circumstance," continues the journal, "produced at the time such a passion for the famous law of offense that anyone

* *Krasnaya Nov*, No. 2.

daring to come out with any sort of reservation in favor of defense would have met with a very poor reception. In order to be a good student of the General Staff it was necessary, even if insufficient, to conjugate interminably the verb *to attack*."

The conservative newspaper *Journal des Debats* for October 5, 1921 launches a sharp criticism from the same standpoint upon the statutes on infantry maneuvers which were published this summer. This newspaper says:

"This splendid booklet begins with an exposition of a whole number of principles which are set forth as the official military doctrine for the year 1921. These principles are admirable; *but why do the compilers continue to pay tribute to an old custom, why do they devote the first page to extolling the offensive?* Why do they advance most prominently to the fore the following axiom: 'He who attacks first exercises an effect upon the psychology of the opponent by revealing a will much stronger than the will of the latter'?"

Having analyzed the experience of two outstanding moments of struggle at the French front, the newspaper then says:

"The offensive can have an effect only on the psychology of an opponent bereft of resources or one that is weak to such a degree as can never be taken for granted. On an opponent conscious of his own strength, the attack does not at all produce an oppressive effect. He does not at all take the enemy's offensive as the manifestation of a will stronger than his own. If the defense has been consciously thought out and prepared as was the case in August 1914 (by the Germans) or in July 1918 (by the French), then, on the contrary, the defensive side considers that its will is the stronger because *the opponent is falling into a trap.*"

The military critic continues:

"You are committing a strange psychological error in fearing the passivity of the Frenchman and his infatuation for defense. The Frenchman is always ready to rush into an offensive, whether he attacks first or second—an offensive that is properly organized. But do not tell him any more Arabian fairy tales about a gentleman who attacks first, being possessed of a greater will."

"The mere fact of attack does not assure success. An attack leads to success when gathered for it are all possible resources which surpass the resources of the opponent. *For in the last analysis he always conquers who proves to be the stronger at the moment of combat.*"

An attempt can of course be made to reject this conclusion on the ground that it flows from the experience of positional warfare. As a matter of fact it flows from maneuverist warfare

with even greater immediacy and obviousness, although in a
somewhat different form. Maneuverist war is war of great
spaces. In the attempt to destroy the enemy's living forces it
does not place value on space. Its mobility is expressed not
only during the offensive but also in retreat, which is only a
shift of position.

11. AGGRESSIVENESS, INITIATIVE AND ENERGY

During the first period of the revolution the Red troops
generally shunned the attack, preferring to fraternize and discuss. During the period when the revolutionary idea was spon-
taneously flooding the country, this method proved very effective. The Whites at that time tried, on the contrary, to force
attacks in order to preserve their troops from revolutionary
disintegration. Even after discussions ceased to be the most
important resource of revolutionary strategy the Whites continued to be distinguished by an aggressiveness greater than
ours. Only gradually did the Red troops acquire energy and
confidence which secure the possibility of decisive actions. The
subsequent operations of the Red Army are characterized in
the extreme degree by maneuverability. Cavalry raids are the
most graphic expression of this maneuverability. However,
these raids were taught us by Mamontov. From the Whites
we likewise learned how to make sudden break-throughs, envel-
oping operations, penetrations into the rear of the enemy. Let
us recall! In the initial period we thought to defend Soviet
Russia against the White detachments by means of cordons, by
holding on to each other. Only later on, having learned from
the enemy, did we close our ranks into a fist and endow these
fists with mobility; only later on did we place workers on
horses and learn to execute large-scale cavalry raids. A little
exertion of our memories already suffices to make clear how
ungrounded, one-sided, and theoretically and practically false
is the "doctrine" alleging that a maneuverist aggressive strategy is peculiar to a revolutionary army as such. In certain
circumstances this corresponds most with a counter-revolutionary army which is compelled to make up for the lack of
numbers by the activity of skilled cadres.

It is precisely in maneuverist warfare that the distinction
between defense and offense is obliterated. Maneuverist war
is a war of movement. The goal of movement is the destruction

of the enemy's living forces, at a remove of 100 versts or so.
The maneuver promises victory if it preserves the initiative in
our hands. The fundamental traits of maneuverist strategy are
initiative and energy and not formal aggressiveness.

The idea that at each given moment the Red Army resolutely took the offensive on the most important front while
temporarily weakening itself on all other fronts; and that just
this characterizes most graphically the Red Army's strategy
during the civil war (see Comrade Varin's article) is correct
in essence, but it is expressed one-sidedly and therefore does not
provide all the necessary conclusions. While *assuming the
offensive* on one front, considered by us at a given moment as
the most important for political or military reasons, we weak-
ened ourselves on other fronts, considering it possible to take
the defensive and *to retreat* there. But, after all, this testifies
precisely to the fact—which strangely enough is overlooked!—
that into our general operational plans retreat entered as an
indispensable link side by side with attack. The fronts on
which we assumed the defensive and retreated were only segments of our general circular front. On these segments there
fought the sections of the one and the same Red Army, its
fighters and its commanders. And if all strategy is reducible
to offense then it is self-evident that the troops on those fronts
where we confined ourselves to defense and even to retreats
must have been subjected to depression and demoralization.
Into the work of educating troops there must obviously enter
the idea that retreat is not flight, that there are strategic
retreats required sometimes by the need to preserve the living
forces intact, at other times in order to shorten the front, and
sometimes in order to lure the enemy in deeper, all the more
surely to crush him. And if a strategical retreat is legitimate,
then it is incorrect to reduce all strategy to offense. This is
especially clear and incontestable, let me repeat, precisely with
regard to maneuverist strategy. A maneuver is obviously a
complex combination of movements and blows, shifts of forces,
marches and battles—with the ultimate aim of crushing the
enemy. But if strategic retreat is excluded from the maneuver
then obviously strategy will acquire an extremely unilateral
character, that is, it will cease to be maneuverist.

12. NOSTALGIA FOR STABLE SCHEMAS

"What kind of army and for what tasks are we preparing?" asks Comrade Solomin. "In other words: What enemies threaten us and through what strategic paths (defense or offense) will we most quickly and economically cope with them?" (*"Military Science and Revolution"* No. 1, page 19.)

Such a formulation of the question testifies most vividly that the thought of Solomin, the herald of a new military doctrine, is completely the captive of the methods and prejudices of old doctrinairism. The Austro-Hungarian General Staff (like many others) elaborated in the course of decades variants of war: variant "I" (against Italy); variant "R" (against Russia), along with corresponding combinations of these variants. In these variants the numerical strength of Italian and Russian troops, their armament, the conditions of mobilization, the strategic concentrations and deployments constituted magnitudes which were stable, if not constant. In this way, the Austro-Hungarian "military doctrine" basing itself on specific political suppositions was firm in its knowledge of what enemies threatened the empire of the Habsburgs, and from one year to the next it pondered how to cope with the enemy most "economically". The thought of the members of the General Staffs of all countries ran in the fixed channels of "variants." The invention of improved armor plate by the future enemy was countered by strengthening the firepower of artillery and vice versa. Routinists educated in the spirit of these traditions would feel themselves quite out of place under the conditions of our military construction. "What enemies threaten us?"—that is, where are our General Staff variants of future wars? And through what strategic paths (defense or offense) are we preparing to realize the variants outlined in advance? Reading the article of Solomin I was involuntarily reminded of the comic figure of the lecturer on military doctrine, General Borisov of the General Staff. No matter what question was under discussion Borisov would invariably raise his two fingers in order to take the opportunity to say:

"This question can be decided only in conjunction with other questions of military doctrine, and for this reason it is first of all necessary to institute the post of Chief of General Staff."

From the womb of this Chief of General Staff, the tree of

military doctrine would spring up and produce all the necessary fruits, approximately in the ancient manner of the fabled daughter of an oriental king. Solomin like Borisov yearns essentially for the lost paradise of the stable schemas of "military doctrine", when it was known ten and twenty years in advance who the enemies were, and how and whence they threatened. Solomin like Borisov needs a universal Chief of General Staff who would gather together the broken pieces of crockery, glue them together, put them on the shelf and paste labels on them: variant "I," variant "R," etc., etc. Perhaps Solomin could at the same time mention to us the universal mind he has in view? So far as we are concerned, we know —alas!—of no such mind and are even of the opinion that there can't be such a mind because the tasks set for it are unrealizable. Talking at every step about revolutionary wars and revolutionary strategy, Solomin has overlooked just this: *the revolutionary character of the present epoch*, which has brought about the complete disruption of stability both in international and internal relations. Germany no longer exists as a military power. Nevertheless French militarism finds itself compelled to follow feverishly the most insignificant events and changes in Germany's internal life and along her borders: What if Germany suddenly raises several million men? What Germany will do it? Will it perhaps be the Germany of Ludendorf? But maybe such a Germany would provide only the impulse which could prove fatal to the existing rotten semi-equilibrium and clear the road for the Germany of Liebknecht and Luxemburg? How many "variants" must the General Staff have? How many war plans is it necessary to have in order to cope "economically" with all the dangers?

I have in my archives not a few reports, thick and thin and medium-sized, submitted by learned authors who with polite pedagogical patience have explained to us that a self-respecting power must institute definite, regular relations, establish in advance its possible enemies, acquire suitable allies or, at least, neutralize all those that can be neutralized. For—as these reporters explained—it is impossible to prepare for future wars "in the dark;" it is impossible to determine either the numerical strength of the army, or its branches, or their disposition. Under these reports I do not recall seeing the signature of Solomin, but his ideas were there. All the authors, sad to say, were from the school of Borisov.

International orientation, including international-military orientation is more difficult nowadays than in the epoch of the Triple Alliance and the Triple Entente. But there is nothing one can do about it: The epoch of the greatest convulsions in history, both military and revolutionary, has destroyed certain variants and stereotyped patterns. There cannot be any stable, traditional, conservative orientation. Orientation must be vigilant, mobile, and expeditious, or, if you prefer, maneuverist. Expeditious does not mean aggressive, but it does mean strictly corresponding to today's combination of international relations and concentrating the maximum forces on the tasks of today.

Under the existing international conditions orientation demands far greater mental skill than was required for the elaboration of the conservative propositions of military doctrine during the past epoch. And, in addition, this work has to be done on a far broader scale and with the employment of far more scientific methods. The fundamental work in evaluating the international situation and the tasks that flow from it for the proletarian revolution and the Soviet Republic is being fulfilled by the party, by its collective thought; and the directive forms are given this work by the party Congresses and its Central Committee. We have in mind not only the Russian Communist Party but also our international party. Solomin's demands for compiling a catalogue of our enemies and determining whether we shall do the attacking and just whom we shall attack appear so pedantic in comparison to the work of evaluating all the forces of the revolution and counter-revolution as they now exist and evolve that has been accomplished by the latest World Congress of the Communist International! What other "doctrine" do you need?

Comrade Tukhachevsky submitted a proposal to the Communist International that an International General Staff be established and attached to it. This proposal was of course incorrect; it did not correspond to the situation and the tasks formulated by the Congress itself. If it was possible to create the Communist International only after strong Communist organizations were formed in the most important countries, then this holds even more for an International General Staff which can arise only on the basis of national general staffs of *several* proletarian states. So long as this remains lacking an International Staff would inescapably be transformed into a caricature. Tukhachevsky found it necessary to deepen his error by publishing his letter at the end of his interesting book, "The War of the Classes." This error pertains to the same order as the headlong theoretical onslaught launched by Comrade Tukha-chevsky against the formation of militia which he alleges, stands in contradiction to the Third International. Let us note in passing that the tendency to attack without the proper safeguards generally consitutes the weak side of Comrade Tukhachevsky, one of the most gifted of our young military workers.

But even without an International Staff which does not correspond to the situation and which is therefore speculative, the World Congress itself, as the representative of revolutionary proletarian parties did accomplish—and through its Executive Committee continues to accomplish—the fundamental ideological work of the "General Staff" of the world revolution: keeping a tally of friends and enemies, neutralizing the vacillators with a view to later attracting them to the side of the revolution, evaluating the changing situation, determining the urgent tasks and concentrating efforts on a world scale upon these tasks.

The conclusions which derive from this orientation are very complex. They cannot be squeezed into a few General Staff variants. But such is the character of our epoch. The superiority of our orientation consists precisely in this, that it corresponds to the character of the epoch and its relations. In accordance with this orientation we align ourselves in our military policy as well. At the present time it is actively vigilant, defensive and preparatory in character. We are above all concerned in assuring with regard to our military ideology, our methods and our apparatus a flexibility so strong as to enable us at each turn of events to concentrate our main forces in the main direction.

13. THE SPIRIT OF DEFENSE AND THE SPIRIT OF OFFENSE

But, after all, Solomin objects "it is impossible at one and the same time to educate in the spirit of defense and in the spirit of offense." (*Loc. cit.*, page 22.) Now, this is sheer doctrinairism. Why can't this be done? Who said that it can't be done? Where and by whom is this proved? By no one and nowhere, for it is false to the core. The entire art of our military construction (and not only military construction) in

Soviet Russia consists in combining the international revolutionary-offensive tendencies of the proletarian vanguard with the revolutionary-defensive tendencies of the peasant masses and even of broad circles of the working class itself. This combination corresponds to the entire international situation. By explaining its significance to the advanced elements in the Red Army we thereby teach them to combine defense with offense correctly not only in the strategical but also in the revolutionary-historical sense of the word. Does Solomin think perhaps that this tends to extinguish "spirit"? Both he and his co-thinkers hint at this. But this is already simon-pure Left SR'ism! The clarification of the essence of the international and domestic situation and an active, "maneuverist" adaptation to it cannot serve to extinguish spirit, but only to temper it.

Or is it perhaps impossible in a *purely military* sense to prepare the army both for defense and offense? But that, too, is nonsense. In his book Tukhachevsky underscores the idea that it is excluded, or almost excluded for the defense in civil war to assume positional stability. From this Tukhachevsky draws the correct conclusion that under such conditions defense, like offense, must of necessity be active and maneuverist in character. If we happen to be too weak for attack, then we strive to detach ourselves from the embraces of the enemy in order later to gather ourselves into a fist and to strike at the enemy's most vulnerable spot. Erroneous to the point of absurdity is Solomin's assertion to the effect that the army is moulded for a specialty—either for defense or for offense. In reality the army is educated and trained for combat and conquest. Defense and offense enter as variable moments into the combat, all the more a maneuverist combat. He conquers who is able to defend himself well when it is necessary to be on the defensive and to attack well when it is necessary to attack. This is the only healthy training which we are obliged to give our army, first and foremost, in the person of its commanding staff. The rifle with bayonet is good both for defense and offense. The same thing applies to the fighter's hands. The fighter himself and the branch of the army to which he belongs must be prepared for combat, for self-defense, for resisting the enemy, for annihilating the enemy. That regiment is best able to attack which is best able to defend itself. Good defense is accessible only to a regiment that has the desire and ability to attack. The statutes must teach how to fight, and not incite to attack.

Revolutionism is a spiritual state and not a ready-made answer to all questions. It can give enthusiasm, it can assure zeal. Enthusiasm and zeal are the most valuable conditions for success but not the only ones. Orientation is indispensable; training is indispensable. And as for doctrinaire blinders—away with them!

14. THE IMMEDIATE TASKS AHEAD

But aren't there in the complex intermeshing of international relations certain outstanding, clear and definitive elements in accordance with which we ought to align ourselves in our military work in the course of the next few months?

There are such elements and they speak far too loudly for themselves to be considered secrets. In the West there are Poland and Rumania; behind their back stands France. In the Far East there is Japan. Around and close to the Caucusus —England. I shall dwell here only on the question of Poland, as the clearest and most intelligible.

Briand, Minister-President of France, has announced in Washington that we are presumably preparing to attack Poland in the spring. Not only every commander and Red soldier but every worker and peasant in our country knows that this is unadulterated balderdash. Briand himself of course knows it, too. Up to now we have paid so big a price to the big and little bandits to get them to leave us in peace, even if temporarily, that it is possible to talk about any "plan" on our part to assault Poland only as a cover for some fiendish plot. What is our actual orientation with regard to Poland?

We are proving to the Polish popular masses firmly and persistently not in words but in deeds—and first of all by the strictest fulfillment of the Riga treaty—that we want peace and are helping in this way to preserve it.

Should the Polish military clique, incited by the French stock-market clique, nevertheless descend upon us in the spring, the war will be on our side both in essence and in popular consciousness, genuinely defensive in character. Precisely this clear and definitive consciousness of our rightness in a war foisted upon us will act to weld together all the elements in the army most closely: the advanced worker–Communist as well as the specialist who is non-party but who is devoted to the Red Army as well as the backward peasant-soldier; and thereby best prepare our army for the initiatory and self-sacri-

ficing offense in this defensive war. Whoever takes this policy
to be indefinite and conditional; whoever remains unclear concerning "what kind of army and for what tasks we are preparing"; whoever thinks that it is "impossible at one and the
same time to educate in the spirit of defense and in the spirit
of offense"—understands nothing at all, and would best keep
quiet and not hinder others!

But if such a complex combination of factors is to be observed in the world situation, then how can we nevertheless
orient ourselves practically in our military construction? What
should be the numerical strength of the army? In what sort
of units? With what dislocations?

All these questions do not permit of any absolute solution.
It is possible to speak only of empirical approximations and
of timely rectifications, depending upon changes in the situation. Only hopeless doctrinaires believe that answers to questions of mobilization, formation, training, education, strategy
and tactics can be obtained deductively, in a formal logical
manner from the premises of a sacred "military doctrine."
What we lack are not magical, all-saving military formulas but
a more careful, attentive, precise, vigilant and conscientious
work resting on those foundations which we have already
firmly lodged. Our statutes, our programs, our army formations are imperfect. This is unquestionable. There is an over-
abundance of omissions, misstatements, inclusions of things that
are outlived, and of others that are incomplete. It is necessary
to correct, improve, render more precise. But how and from
what standpoint should this be done?

We are told that it is necessary to put the doctrine of
offensive warfare as the basis for review and rectification.
Solomin writes:

> "This formula signifies the most decisive (!) turn (in the
> construction of the Red Army); it is necessary to review all
> (!) the opinions we now hold, to carry out a complete (!) re-
> evaluation of values from the standpoint of passing over from
> the purely defensive strategy to that of offense. The education
> of the commanding staff, the preparation of the individual
> fighter . . . armament—all this (!) must henceforth proceed
> under the sign of offense . . ." (*Loc. Cit.*, page 22.)

He also writes:

> "Only if such a single plan is given will the reorganization
> of the Red Army, which has already begun, emerge from a
> condition of formlessness, dispersion, disharmony, vacillation
> and the absence of a clearly conscious goal."

Solomin's language is, as we see, rigidly aggressive, but his assertions are absurd. Formlessness, vacillation and dispersion exist in his own mind. Objectively, our work contains difficulties and practical mistakes. But there is no dispersion, no vacillation, no disharmony. The army will not permit the Solomins to incorporate their vaporings on organizational and strategical matters and in this way introduce vacillation and dispersion.

Our statutes and programs must be reviewed not from the standpoint of the doctrinaire formula of pure offensive but from the standpoint of our experience of the last four years. It is necessary to read, discuss and check the statutes at conferences of our commanding personnel. It is necessary to juxtapose the still fresh recollections of combat actions, major and minor alike, with the formulations in the statutes; and each commander must consciously ask himself whether or not the words correspond to the deeds, and if not, just where do they diverge. To gather this organized experience, draw its balance sheet, appraise it in the center by applying the strategic, tactical, organizational, political criteria of experience of a higher order; to cleanse our statutes and programs of everything that is outlived and superfluous; to bring them closer to the army and to instill in the army the feeling of how indispensable they are and to what measure they can replace crude handiwork—here is really the big, urgent immediate task!

We possess an orientation international in its scope and of a great historical sweep. One of its sections has already passed the test of experience; another section is now being verified and is meeting the test. The Communist vanguard is sufficiently assured of revolutionary initiative and aggressive spirit. We do not need verbal, noisy innovations with regard to new military doctrines, nor proclamations of them with the beating of drums; what we need is the systematization of experience, improvement of organization, attention to little details.

The gaps in our organization, our backwardness and poverty, especially in the field of technology, must not be erected by us into a *credo*. Instead we must do everything in our power to eliminate them, seeking to approach in this respect the imperialist armies which all deserve to be crushed but which nevertheless possess certain superiorities: rich aviation, abundant means of communication, well trained and carefully selected commanding personnel, precision in calculating resources, maintenance of correct reciprocal relations. This is of course only an organization-technical integument. Morally and politically the bourgeois armies are disintegrating or heading toward disintegration. The revolutionary character of our army, the class homogeneity of our commanding personnel and of the mass of the fighters, the Communist leadership—here is where our most powerful and unconquerable force lies. None can take it away from us. All our attention must now be directed not toward a fantastic reconstruction but toward improvement and greater precision. To supply sections properly with food; not to permit products to rot; to cook good cabbage soup; to teach how to destroy body vermin and to keep clean; to correctly conduct training exercises, doing it less within four walls and more under the open sky; to prepare political discussions intelligently and concretely; to furnish each Red soldier with a service book and keep good records; to teach how to oil rifles and grease boots; to teach how to shoot; to help the commanding personnel thoroughly assimilate the statute regulations concerning maintaining communications, gathering intelligence, making reports, maintaining guards; to learn and to teach how to adapt oneself to various localities; to wrap one's feet correctly in pieces of cloth to keep them from getting rubbed raw; once again to grease boots—such is our program for the next winter and the coming spring.

Should anyone, on a holiday occasion, choose to call this practical program a military doctrine, he will not be held to account.

3. OUR CURRENT BASIC MILITARY TASKS

April 1, 1922

TROTSKY'S REPORT

I

What Are We Discussing?

First, a few preliminary remarks relating to the history of the question before us. A critical and impatient movement in favor of a new military doctrine manifested itself even before the Tenth Party Convention. The Ukraine was the chief breeding ground of this movement. More than a year ago Comrades Frunze and Gussev formulated theses devoted to a unified military doctrine, and tried to get them adopted by the Convention. In my capacity as reporter on the Red Army question I declared at the time that these theses were in my opinion false from the standpoint of theory and fruitless from the standpoint of practice. Comrades Frunze and Gussev then withdrew their theses which, of course, does not at all mean that my arguments had convinced them. Among those engaged in military work there has continued to exist a certain grouping under the banner of a "proletarian military doctrine." It is only necessary for you to recall the article of Comrade Solomin, certain speeches of Comrade Gussev, and so on. I felt myself obliged to relinquish my previous position of watchful waiting inasmuch as the articles by

Solomin and others might, if permitted to pass unchallenged, sow the greatest confusion in the minds of the army's leading elements. There has been no answer as yet to my article, *Military Doctrine or Pseudo-Military Doctrinairism*. Nevertheless differences of opinion and prejudices on this question have not been outlived, although there is no longer any room for doubt that on this subject the public opinion of the overwhelming majority of the party has already become fixed.

The task of the present discussion which has been initiated by Comrades Frunze and Voroshilov is to clarify this same question of military doctrine. The external impulsion for the discussion came from the programmatic theses on training and educating the Red Army, defended by Comrade Frunze at the recent conference of the Ukrainian commanders. I must begin by saying bluntly that these theses are in my opinion more dangerous and harmful than the articles by Comrade Gussev and others on the same subject. Comrade Solomin's article runs far too obviously counter to the logic of things, counter to common sense and counter to our own experience. It was obviously written in a moment of doctrinaire occultation. I am sorry that the author is not here and unable to defend his viewpoint personally. But his article is a political fact and I am constrained to deal with it lest it exert further harmful influence. As regards the Ukrainian theses, they are far more cautiously worded, and so combed and cleaned that at first glance everything appears to be in good order; more than that —and here I must render to the author of the theses what is due him for his artistry in maneuvering—certain points are accompanied with a notation in parentheses: Trotsky, Trotsky, Trotsky . . . An impression is created that these might almost be actual quotations from my articles. The terminology has likewise been renovated. The word "doctrine" has been supplanted by the expression "unified military world-outlook," which is, in my opinion, 100 times worse. And here we already pass from the history of the issue to its essence.

A unified military doctrine obviously presupposes that we likewise have a unified industrial doctrine, a unified commercial doctrine, etc., so that from the sum-total of these there arises a unified doctrine of Soviet activity. This is a pompous and an affected terminology, but still tolerable. But by writing "unified military world-outlook," the point is driven home far more strongly. It now turns out that there exists some sort of "military" *outlook upon the entire world.* Up to now we have proceeded on the assumption that we have a Marxist world-outlook. And we suddenly hear that it is also necessary to have a unified military world-outlook. No, Comrades, get rid of this terminology as quickly as possible!

In polemicizing against the term "doctrine," I disclaimed any intention of starting a fight over a word. But, in my opinion, the totality of views and moods for which this term serves as a cover, is very dangerous.

Let us get down to cases. The theses tell us that a unified military world-outlook represents a totality of views, raised to a system with the aid of the *Marxist method* of analyzing social events. Here is how point 1 reads verbatim:

> "This education and training permeating all the stratifica-tions of the army must be carried out on the basis of unified views on the fundamental questions relating to the tasks of the Red Army, the elementary principles of building it, and the methods of conducting combat operations. It is precisely the totality of these views raised to a system with the aid of the Marxist method of analyzing social events and inculcated in the Red Army through statutes, orders and regulations that provides the army with the necessary unity of will and thought."

The Trade of War and—Marxism

Does this include strategy, tactics, military technology and our military statutes? Are these included in the "totality of views raised to a system with the aid of the Marxist method"? Yes or no? It is necessary to answer this question. In my opinion, they must be included. How can it be otherwise? After all, statutes—not in the sense of our statute booklets but in the sense of their principles—must enter into this "unified world-outlook," mustn't they? For once they are thrown out, nothing military remains. In that case one is simply left with a "world-outlook." What determines its military character are precisely the statutes which sum up military experience and which determine our military usages. But have our statutes then been created by means of the Marxist method? This is the first time I hear of it. Statutes sum up military experience. Our statutes may perhaps limp, and we shall continue to per-fect them on the basis of our military experience. But how can they be unified by means of the Marxist method?

What is the Marxist method? It is a method of thinking scientifically. It is the method of historical social science.

True enough, our army magazine bears the name: *Military Science*. But our magazine still contains many incongruities left over from the past, and most incongruous of all is its name. There is not and there never has been a military "science." There does exist a whole number of sciences upon which military affairs rest. Included among them essentially are all the sciences from geography to psychology: An outstanding army leader must possess the knowledge of the elementary principles of many sciences—although, to be sure, there are self-taught army leaders who act on the basis of probing empirically, but who are assisted by a certain innate sense. War rests on many sciences, but war itself is not a science—it is a practical art, a skill. The Prussian strategist, King Frederick II was fond of saying that war is a trade for an ignoramus, an art for a man of talent and a science for a genius. But he told a lie. This is false. For an ignoramus war is not a trade because ignorant soldiers are the cannon fodder of war and not at all its "tradesmen." As is well known, each trade requires a certain schooling; and for those who are correctly schooled in military affairs war is therefore a "trade." It is a cruel, sanguinary trade, but a trade nonetheless, that is, a skill with certain habits which are elaborated by experience and correctly assimilated. For gifted people and those of genius, this skill becomes transformed into a high art.

War cannot be turned into a science because of its very nature, no more than it is possible to turn architecture, commerce or a veterinary's occupation into a science. What is commonly called the theory of war or military science represents not a totality of scientific laws explaining objective events but an aggregate of practical usages, methods of adaptation and proficiencies corresponding to a specific task: the task of crushing the enemy. Whoever masters these usages to a high degree and on a broad scale and is able to attain great results by means of combinations—such an individual raises military affairs to the level of a cruel and sanguinary *art*. But there is no ground whatever to talk of science here. Our statutes are just a compilation of the practical rules derived from experience.

The Quagmire of Scholasticism

Marxism on the other hand is a method of science, that is, the science of apprehending objective events in their objective connections. Just how is it possible to construct the usages of a military trade or art by means of the Marxist method? This is the same thing as trying to construct a theory of architecture or a text book on veterinary medicine with the aid of the Marxist method. A history of war, like a history of architecture can be written from the Marxist viewpoint, because history is a science. But a so-called theory of war, i.e., practical [military] leadership is something else again. These must not be mixed up, otherwise the result is not a unified world-outlook but the greatest muddle.

With the aid of the Marxist method social-political and international orientation is facilitated in the extreme. This is incontestable. Only with the aid of Marxism is it possible to analyze the world situation, especially in our modern and exceptional epoch.

But it is impossible to construct a field statute with the aid of Marxism. The blunder here lies in interpreting military doctrine or, what is worse, "unified military world-outlook" to include our general state orientation, both international and internal, as well as practical military usages, statute regulations and precepts—with the expressed desire of seemingly rebuilding all this anew with the aid of the Marxist method. But our state orientation has long been built and is still being built by means of the Marxist method and there is no need whatever of starting to build it anew within the womb of the war department. With regard to the purely military methods —as they are set down in our statutes—it is hardly expedient to apply the Marxist method here. It is of course necessary to introduce the maximum of unity into the statutes and check them against experience, but it is sheer absurdity to talk about the unified military world outlook in this connection.

Such are the first and second points of Comrade Frunze's theses.

I now come to point 3:

"The elaboration of this unified world-outlook of the workers' and peasants' army was started at the very first stages of its existence."

This almost seems a polemic against Comrade Gussev who has given us to understand that we never had and still haven't got any principles of construction.

"In the course of further practical work were crystallized and delineated all the basic elements of the military system of the proletarian state, which flow from its specific class nature."

This takes in far too much territory. It turns out that our military system derives wholly from the specific class nature of the proletarian state. Presumably the task is first to determine this nature, next deduce from it a unified military doctrine, and then obtain from the latter all the necessary partial, practical conclusions. This method is scholastic and hopeless. The class nature of the proletarian state determines the social composition of the Red Army and particularly of its leading apparatus; it determines its political world-outlook, its aims and its moods. Naturally, all this exerts a certain indirect influence upon strategy and tactics alike, and yet strategy and tactics are not derived from a proletarian world-outlook but from the conditions of technology, in particular military technology, from the available facilities of providing supplies, from the geographical milieu, the character of the enemy, etc., etc.

Do we possess a unified industrial or a unified commercial world-outlook? Is it possible for us to deduce from the "specific nature of the proletarian state" the best textbook of foreign trade, or the best method of administrative or commercial organization for our trusts? An attempt to do this would be ludicrous and hopeless. To think that by arming oneself with the Marxist method it is possible to solve the question of how best to organize production in a candle factory, is to understand nothing either about the Marxist method or about a candle factory. Meanwhile, an army regiment from the standpoint of its own specific tasks is a factory that must be correctly organized, that is, in harmony with its purposes. I assert that an attempt to derive from the system of the proletarian state by means of deduction, i.e. logically, the organization, structure, and tactical usages of an infantry or cavalry regiment is absolutely utopian and nonsensical.

The authors of the criticized theses themselves sense this because they keep wavering between the "unified proletarian doctrine" and the French field statutes for the year 1921. I shall deal with this later on.

No Abstractions—Only Concretizations!

The premises for the existence of an army are of course wholly political in character. The state must have an answer to the question: What kind of army are we preparing and for what tasks? But inasmuch as our army is revolutionary and class-conscious it must itself also have a clear and correct answer to this question. Point 4 of the Ukrainian theses sets this as its aim. I consider it to be politically one of the most dangerous passages. In it we read the following:

> "The profound principled contradiction between the system of proletarian state-ism on the one hand and the surrounding bourgeois capitalist world on the other renders inevitable both conflicts and a struggle between these two hostile worlds. In correspondence with this, the task of educating the Red Army politically consists in reinforcing and strengthening its constant readiness to engage in a struggle with world capitalism. This combat mood must be riveted by means of planful political work, carried out on the basis of proletarian class ideology, in forms that are viable and accessible to all."

Such an approach to the question is patently non-political, abstract, wrong and dangerous in its essence. The struggle between the proletariat and the bourgeoisie is being waged throughout the whole world. In the course of this struggle either our country will be attacked or we shall do the attacking. The army must be held in readiness, educated on the basis of proletarian class ideology—"in forms that are viable and accessible to all." But all this is the most abstract communist doctrinairism against which all of us made speeches during the previous session when we discussed military propaganda! A beautiful program is offered us: in the first part of the year convert one-quarter of the peasant Red soldiers into communists; in the second part add another quarter, and then still another, and in this way, that is by means of barracks propaganda, alter the reciprocal relations of classes within the country and create an army which would proceed in its political consciousness from the international proletarian class ideology as its motive force. But such an approach is false to the core, patently utopian.

Yesterday we were all seemingly saying: Don't forget that our army in its overwhelming majority consists of young peasants. It represents a bloc between the directing worker minority and the peasant majority led by it. The basis of the bloc is the need of *defending* the Soviet Republic. It must be defended because it is being attacked by the bourgeoisie and the landlords—foreign and domestic enemies.

The entire strength of the workers' and peasants' bloc rests upon the conscious recognition of this fact. Naturally, we reserve the programmatic right of dealing blows to the class enemy on our own initiative. But our revolutionary right is one thing, the realities of today's situation and of tomorrow's perspectives are something else again. Some may take this to be a secondary distinction, but I assert that the life and death of our army depends on this. Those who do not understand this, understand nothing about our entire epoch and, in particular, they do not understand what the NEP is. It is as if we said: On the basis of proletarian ideology—"in forms that are viable and accessible to all"—the entire people must be educated in the spirit of the socialist organization of economy. This is easily said! But in that case what need is there for a new economic policy [the NEP] with its decentralization, its market, etc.? This, it will be said, is a concession to the moujik. Yes, it is just that. Failing this concession, the Soviet Republic would be overthrown. How many years will this economic zone endure? We don't know—it may be two years, three years, five or ten before the revolution comes in Europe. How do you want to get around this with your "military world-outlook"? You want the peasant, on the basis of proletarian doctrine, to be prepared at any moment to wage war on the international fronts for the cause of the working class. It is our direct duty to educate communists, advanced workers in this spirit. But to think that an army, an armed bloc of workers and peasants, can be built on this basis—is to be a doctrinaire and a political metaphysician, because the peasantry becomes imbued with the idea of the necessity of maintaining the Red Army only to the extent that it becomes convinced that despite our intense efforts to preserve peace and despite our greatest concessions, the enemies continue to threaten our existence.

Naturally, the situation may change: Great events in Europe can create entirely different conditions for military initiative on our part. This is in complete harmony with our program. But, after all, you are not engaged in writing a program. We have to elaborate methods of educational work for the present day and not for eternity. And the basic decisive slogan which corresponds to the entire situation and to our entire policy is *defense*. In the era when the army is being demobilized on a vast scale and when it is being constantly reduced, in the era of the NEP, in the era of the preparatory organizational and educational work in the European proletarian movement—after the already executed retreats—in the era of the working class united front, that is at the same time when joint practical action with the Second and 2½ Internationals is being attempted—in this era it is ludicrous and absurd to say to the army, "It may be that the bourgeoisie will assail us tomorrow or it may be that on the morrow we will attack the bourgeoisie."

To do so is to distort the perspectives, to befuddle the minds of the Red soldiers so as to make it impossible for them to grasp the educational significance of our international spirit of conciliatoriness, and to paralyze the great educational, revolutionary force of this conciliatory policy which will manifest itself in the event that we are attacked despite all our efforts.

The "Concession" to the Red Soldier-Peasant

It might seem that all these considerations have been amply clarified both within our party and on an international scale; the Third World Congress and the recent party conferences were largely devoted to these questions. But no sooner do we set ourselves the task of creating some sort of unified military world-outlook, than all the established political premises for our internal and international work are flung to the winds and we take naked abstractions as our starting point: "The international class struggle . . . we are being attacked . . . we shall do the attacking, etc. . . . . we must be prepared to take the offensive . . ."

It is impossible to carry out with impunity an experiment of this sort with the consciousness of the Red Army mass. The army mass wants to know and, together with all the toilers of our country, has the right to know: What kind of army are we preparing and for what tasks? Not for the year 1930, but today. Why must we remain [in the army] under the banner of 1899, and for how long? Our answers to these questions will be clear and convincing, only if we do not begin by mixing ourselves up.

But Point 5 multiplies this doctrinaire blunder. It states flatly that "*the army will henceforth fulfill its combat tasks under the conditions of revolutionary war, either defending itself against the onslaught of imperialism or advancing shoulder to shoulder with the toilers of other countries in a joint struggle.*" These two eventualities are juxtaposed as if they were equally applicable to today: it is a case of either the one or the other. Well, just how will you tell a Saratov peasant: We shall either lead you to Belgium to overthrow the bourgeoisie, or on the other hand, you will have to defend Saratov *goubernia* [province] against an Anglo-French expeditionary force in Odessa or Archangel? Could one pose the question in this way without biting his tongue? Of course not! In speaking before a regiment or before a meeting of workers and peasants, each of us would invariably draw close to reality and say: We are prepared, under such and such terms, to pay Czarist debts because we wish to avoid war; but our very powerful enemies are engaged in machinations. We are still compelled to retain the status of the year 1899 within our army . . .

The more factually, the more concretely we present to our audience the difficulties of our international position, the magnitude of our concessions, all the more clearly will they be able to grasp the need of preserving the Red Army, and, at the same time, all the more will our words correspond to the truth of today. But if we advance a "doctrine" of either being ourselves attacked, or ourselves doing the attacking—then we can only introduce confusion into the minds of our commissars, political directors and commanders, for we shall have given them a false picture of reality, and invested the entire agitation with a false tone. With such abstract speeches we can never reach the moujik's heart. It is the surest way of ruining our military propaganda and our political agitation.

An Attempt at Philosophy

I now come to the sixth point of the theses. Here we pass from politics to strategy, that is, into the sphere of purely military questions. The theses, as you know, were formulated by Comrade Frunze. I must avoid any possible misunderstandings, I must say that I esteem Comrade Frunze as one of the most gifted of our military workers, and I would never undertake myself the practical strategical work with which I would entrust him. But under discussion now is not Comrade Frunze's work as an outstanding army leader but his attempt to create a military philosophy. The late Plekhanov who towards the end of his life committed many sins in politics was, as is well known, extremely exacting in questions of philosophy. Plekhanov used to say that a Marxist has the right not to study philosophy—but if you are the kind of person who does take it up, and even out loud, then don't muddle. This was his favorite precept. Wherever he caught anyone in philosophical deviations he would attack like a wolfhound. Sometimes he was told: "George Valentinovich, why do you attack so cruelly? Perhaps the poor fellow hasn't even had the time to study philosophy." And Plekhanov would answer: "Then let him hold his peace and not spout 'independent' notions of his own because the most harmful political

consequences can result from this." Plekhanov caught up
Peter Struve on his muddling in philosophy long before Struve
began to stray from Marxism politically.

We have before us here not philosophy in the strict sense
of the word but rather an attempt at military philosophy. We
are not at all obliged to take up such studies now. We have
a general orientation. In military affairs one can be an
empiricist, correcting and setting things straight on the basis
of experience. In the sphere of military-organizational work
I have taken the liberty of proceeding empirically and would
take no exceptions whatever if Comrade Frunze chose to
remain an empiricist in the strategical field. But he has occupied himself with generalizations and has passed over into the
field of the philosophy of strategy, and in my opinion he has
muddled up things. His own strategic roots are very strong,
but he can cause others to stray from the correct road.

Here is how point 6 reads:

> "Up to now our revolution has had to wage its struggle by
> employing the same basic methods of military tactics and
> strategy as those which have been also employed in the armies
> of bourgeois countries."

Please take particular note of this. Now let us read on:

> "But the change in character and in the living forces of the
> Red Army produced by the revolution through the transfer of
> the leading role to the proletarian elements within the army,
> has found its reflection in the character of applying the gen-
> eral usages of tactics and strategy."

This language is ponderous and vague. But let us go on.
In point 7 it is stated:

> "Our civil war was primarily maneuverist in character. This
> came as a result not only of purely objective conditions (the
> vastness of the theatre of military operations, the relative
> numerical strength of the troops, etc.), *but also of the internal
> traits of the Red Army, its revolutionary spirit, its military
> zeal, which are the manifestations of the class nature of its
> leading proletarian element, etc., etc.*"

We have just been told that up to now we based ourselves
on "bourgeois" strategy; but in the next breath it is asserted
that our civil war was maneuverist in character owing to the
class nature of the proletariat. This discrepancy is not acci-
dental. To say that the maneuverist character of the war was
determined not only by material conditions (vast spaces,

sparsity of troops) but also by "internal" traits of the Red Army as such is to make an assertion that is false from beginning to end. There is no basis for it, nor can a basis be supplied for it, and it reeks of braggadocio.

The Traits of Our Maneuverability

We must begin by analyzing our maneuverability. It evolved first among our enemies and not among us—after all, it is an historical fact that our enemies taught us maneuverability. I have already proved this in my article on military doctrine. Infatuation with maneuverability dates back in particular to cavalry raids and, once again, these were initiated by the Whites who executed them in the beginning better than we did. They taught us maneuverability. This is the first and foremost fact. No one can deny it. It flowed from the fact that their troops were more highly skilled, with an officer cadre personnel far larger than ours. In the beginning they had more cavalry (Cossacks!). For this reason they were better adapted to maneuverability. At the same time they had less of the peasant mass, and whatever they did have was for political reasons far less stable than our peasant mass. This made maneuvering indispensable for them. They tried to make up in speed (mobility) what they lacked in mass. We learned from them. This is an incontrovertible fact. Therefore if you say that maneuverability flows from the revolutionary nature of the proletariat then how will you be able to account for the strategy of the Whites? Your contention is glaringly false!

There is one thing that can be said: Maneuverability in the precise sense of the term is inaccessible to the peasantry both in its revolutionary and counter-revolutionary movements. Because when the peasantry is left to its own resources, the truly peasant form of war is guerrilla warfare (similarly in religion the peasantry is unable to go beyond the sect—it cannot create the church). The peasantry is incapable of creating a state with its own forces—we have seen a particularly graphic illustration of this in the case of the Ukrainian Makhnovist movement. In order to lift the peasantry to the level of a state and of an army, the hand of some one else over them is needed. Among the Whites it is the nobility, the landlords and the bourgeois officers who have managed to learn a few things from the landlord-officers. They take the peasants by the throat, place over them a centralized apparatus of coercion, teeming with officers and—proceed to maneuver. Among us the directing role is played by the workers who attract the peasantry, organize it and lead it forward. To the extent that maneuverability (not guerrilla warfare!) presupposes a centralized military organization during the civil war, to that extent maneuverability was peculiar to both camps. Please do not tell us that maneuverability flows from the revolutionary traits of the proletariat. This is not true. It flows from the size of the country, from the numerical strength of the troops, from the objective tasks posed before an army as such and not at all from the revolutionary nature of the proletariat.

And just what have been hitherto the traits of our maneuverability? The basic trait is, alas, formlessness . . . We have many reasons, Comrades, for being proud of our past but we have no right to idealize it uncritically. We must study, we must keep going forward. And for this, it is necessary to know how to appraise critically, but not how to sing hymns.

We Need Not a "Doctrine," but Cadres!

There has been virtually no critical analysis of the maneuverability of the civil war, nor a critical evaluation of it undertaken as yet, and failing this we shall be unable to take a step forward. There were admirable individual plans, there were operations, brilliant in the maneuverist sense, which secured us many victories, but on the whole our strategical line was characterized by formlessness. We attacked stormily and resolutely, we maneuvered audaciously, but not infrequently our maneuver resulted in our having to leap back hundreds of versts. To find an explanation for our maneuvers in the revolutionary character of the proletariat, combat spirit, etc., is to be thinking in a fog. The revolutionary character of the advanced workers and class-conscious peasants finds its expression in their self-abnegation, in their heroism—during all kinds of operations, under all kinds of strategy. Whereas the explanation for the instability and formlessness of our maneuverist strategy lies time and again in the inadequate organization of our zeal for combat: we still lacked real, serious cadres. Herein is the key to the question: our lower commanding staff was too weak, our intermediate commanding staff inadequately trained. That is why our plans, sometimes superb ones, broke down and were atomized in the course of execution and resulted in gigantic leaps backwards. On almost all the fronts we had to fight the war twice, sometimes three times. Why? Because of the quantitative and qualitative deficiencies of the cadres.

War is always an equation with many unknowns. It cannot be otherwise. If all the elements of war were known in advance, then there would be no wars: able to foresee the results in advance one side would simply surrender without battle to the other. But the task of military art does consist in reducing to a minimum the quantity of unknowns in the war equation, and it is possible to achieve this only by assuring the maximum of harmony between a plan and its execution. What does this mean? It means having such military formations and such a leading personnel as would assure the attainment of the goal set through overcoming the obstacles of space and time by means of combined methods. In other words, it is necessary to have a commanding apparatus that is stable and at the same time flexible, that is centralized and at the same time elastic, that has mastered all the necessary habits and is able to pass them on to the ranks. Good cadres are necessary. This question cannot be solved by singing paeans to revolutionary maneuverability. There has been no lack of maneuverability; and still less do we or did we feel any lack of idealization of maneuverability. You could say that if our commanding staff did ail from anything toward the end of the civil war, it was precisely from an excess of maneuverability—from a sort of *maneuverist intoxication*. All the talk was about maneuvers. Cavalry raids were seen in dreams. But what do we actually lack? Stability in the maneuver itself, stability that can be secured only by a good commanding staff of a maneuvering army. This is where our center of attention must be shifted during the coming period of training. The schematic idealization of maneuverability which allegedly flows from the class nature of the proletariat does not lead us forward but keeps us back and even drags us back.

The Danger of the Abstraction of a "Civil War in General"

The idea contained in point 8, as it is expressed here, secretes a danger not only, and not so much, for us as for the revolutionary parties of other countries. It is impermissible to forget that others are now learning from us; and when we occupy ourselves with revolutionary generalizations, including revolutionary-military generalization, we must not only always bear in mind Moscow and Kharkov, but also watch out

lest we sow misunderstandings in the West. Point 8 of the
theses states:

> "The conditions of the future revolutionary wars will pre-
> sent a number of peculiarities which will bring these wars
> closer to the civil war type. In connection with this the
> character of these wars will unquestionably be maneuverist.
> Therefore our commanding staff must be educated primarily
> in the ideas of maneuverability and mobility, while the entire
> Red Army must be prepared and trained in the art of quickly
> and planfully carrying out march-maneuvers."

By revolutionary wars are meant here the wars of the
workers state against bourgeois states, as distinct from a pure
civil war, that is, a war between the proletariat and the bourgeoisie of one and the same state. Point 8 expresses the idea that
future revolutionary wars will approximate civil wars in type,
and for this reason will be maneuverist in character. But just
which civil war is being referred to here? The reference is
obviously to our civil war which took place under the specific
conditions of our immense spaces, sparse population and poor
means of communication. But the misfortune lies in this, that
the theses posit some sort of abstract type of civil war, taking
as their starting point the alleged fact that maneuverability
flows from the class nature of the proletariat and not from the
reciprocal relations between the theater of war and the density
of troops. But, after all, in addition to our civil war, we
know of still another and sufficiently large-scale example in
France—the Paris Commune! In this instance the immediate
task consisted in defending the fortified Parisian place of
arms, from where alone any future offensive could have unfolded. What was the Commune in a military respect? It
was the defense of the fortified Parisian region. Defense could
and should have been active and flexible, but Paris had to be
defended at all costs. To sacrifice Paris for the sake of a
maneuver would have meant to cut down the revolution at its
roots. The Communards were unable to defend Paris; the
counter-revolution conquered Paris and slaughtered tens of
thousands of workers. How then can I, proceeding from the
experience of the steppes of the Don, the Kuban and Siberia,
tell the Parisian worker: From your class nature there flows
maneuverability. A generalization of this sort, hastily made,
is no joking matter!

In the highly developed industrial countries with their dense
populations, with their huge living centers, with their White
Guard cadres prepared in advance, the civil war may assume—

and in many cases will undoubtedly assume—a far less mobile, a far more compact character, that is, one approximating positional warfare. Generally speaking, there cannot even be talk of some sort of absolute positionalism, all the more so in a civil war. In question here is the reciprocal relation between the elements of maneuverability and of positionalism. And here it is possible to state with certainty that even in our supermaneuverist strategy during the civil war the element of positionalism did exist and in certain instances played an important role. There is no room whatever for doubt that *in the civil war in the West the element of positionalism will occupy a far more prominent place than in our civil war.* Let some one try to dispute this. In the civil war in the West the proletariat, owing to its greater numerical strength will play a far greater and more decisive role than in our country. From this alone it is clear how false it is to tie up maneuverability with proletarian class nature. Hungary, during its Soviet days, didn't have sufficient territory to be able to create an army by retreating and maneuvering; for this reason the revolution had to be surrendered to the enemy (interjection by Voroshilov: "They can maneuver in a different way."). Naturally, it is a wonderful idea that it is possible to maneuver "in a different way," that is, to include maneuvers within the framework of defending a given place of arms. But in such a case positionalism would already dominate over maneuverability. Up to a certain point maneuvers will play an auxiliary role during the defense of a given region which is the proletarian hearth of the civil war itself. But when we speak of the maneuverist strategy of civil war what we have in mind is the Russian example wherein we manipulated enormous distances and cities with a view to preserving our living forces and preparing a blow at the living forces of the enemy. During the days of the Commune the situation in France was such that the loss of Paris meant the doom of the revolution. In Soviet Hungary the arena of struggle was larger but it still remained very restricted. But even our arena of maneuverability is not unlimited. We deceive ourselves, not infrequently forgetting that the counter-revolution moved up on us from the border regions which are without any really viable hearths of the revolution. Hence derived the wild sweep of operations and monstrous retreats without mortal danger and without mortal consequences to the Soviet Republic. To the extent that the Whites drew closer to Petrograd, on the one hand, and to Tula, on the other, our place of arms acquired for us an uncondi-

tionally vital significance. We cannot surrender Petrograd or
Tula or Moscow in order later to "maneuver" on the Volga or
the northern Caucasus. Of course, even the defense of the
Moscow place of arms (had our enemies in 1919 scored further
success) would not have necessarily brought us to the immobility of trench warfare. But the need of hanging on to territory and of defending every square verst would have confronted
us far more imperiously. And this means that the elements
of positionalism would have grown enormously at the expense
of the maneuverist elements.

Point 10 of the theses recognizes positionalism—but immediately adds, in holy alarm, that it would be extremely dangerous for us to "permit ourselves to be carried away by positional methods as the basic form of struggle." Why so?
Where did our comrades discover any danger of our being carried away by positionalism? There is intoxication among us,
but it is maneuverist and not at all positional . . . Is the
reference perhaps to our military engineering department
which has recently been building far too many fortresses?
Otherwise this reservation makes no sense at all.

The Proletarian Strategy of . . .
Marshal Foch

Point 11 reads:

"Red Army tactics have been and will remain permeated with
activity in the spirit of offensive operations executed boldly and
energetically. This flows from the class nature of the workers'
and peasants' army (what, again!) and at the same time this
coincides with the requirements of military art."

"This coincides!" What a superb construction! Maneuverability, which flows from the class nature of the proletariat
happens to coincide exactly with the requirements of military
art which had been created by other classes!

"*All other conditions being equal,* the attack is always
more advantageous than the defense." If all other conditions
are equal, then this is correct; there is no gainsaying it. But
this is not all. Further on we read: "Because he who attacks
first exercises an effect upon the psychology of the opponent
by revealing a will much stronger than the will of the latter"
(*French field statutes of 1921*). So, as you see: our strategy
must be offensive, in the first place, because this flows from
the class nature of the proletariat and, secondly, because this

coincides with the French field statutes of the year 1921. *(Laughter. Voroshilov interjects: "There is nothing funny about it.")* No, there is. This reminds me, esteemed Comrade Voroshilov, a little, of the Wuerttemberg democrats of 1848 who used to say: We want a Republic, but with our good Duke at its head . . . So, too, here—we want a genuine proletarian strategy, but one that meets with the approval of Marshal Foch. It seems to be more reliable this way. A Republic, and moreover one headed by a Duke—that is already the best! *(Laughter.)* According to Comrade Voroshilov, there is of course nothing funny about it—but the sooner you delete this, the better it will be for the theoretical merit of our Army.

And besides, it happens to be essentially false. In the first place, this thesis of Foch or somebody else—I do not know who edited the new French field statutes—is now being subjected to a merciless crossfire precisely in French military literature. The offensive is, of course, superior to the defensive. Without the offensive, victory cannot be gained. But to say that he who attacks *first exercises* an effect upon the psychology of the opponent is to fall into offensive formalism. Without the offensive, victory cannot be gained. The offensive is in the last analysis superior to the defensive. But it is not necessary to be invariably the first to attack; the offensive should be launched when it is indicated by the situation.

Were We to Think Concretely . . .

A booklet, *On the Principles of Military Art*, signed with initials X.Y. has recently been published by a French author. German military literature has acclaimed this book as the most remarkable military work in France since the war. The author of this book comes out most emphatically against the thesis of the new French field statutes which has been cited by Comrade Frunze. The author adduces as an illustration the attempt of the French to be the "first" to attack on the Lorraine theater of war in 1914, where the Germans, in their fortified positions, sat calmly awaiting the enemy's offensive. Therewith the moral preponderance was wholly on the side of a calculated and well prepared defense, which happened to be an outright trap for the attacking force. During the last period of the war the Germans assumed the initiative in the summer offensive of 1918. The Anglo-French army, after withstanding the offensive and draining the enemy forces, passed over in its turn from flexible defense to a counter-offensive which proved fatal to the army of Hohenzollern. Without the offensive, victory cannot be gained. But victory is gained by him who attacks when it is necessary to attack and not by him who attacks *first*.

Isn't it time to stop talking about the "offensive in general"? Many people proceed by mentally tearing out of the operations of the civil war some segment wherein we attacked successfully and victoriously; and taking this experience as a starting point, they depict to themselves, along this model, a picture of our future offensives. It is necessary to learn to think more concretely. Those states which may drag us into war are known to us. The possible theater of war is consequently open to scrutiny. War begins with mobilization, concentration and deployment of forces. In our strategic forecasts it is therefore necessary to proceed from the preparatory operations—first of all, mobilization. Who, then, will begin to attack *first?* Obviously, that opponent who is able to gather sufficient forces for it. Does mobilization give us the necessary preponderance? Sad to say, it does not. With the technical aid of imperialist countries our possible opponents may possess a certain preponderance with regard to technology—not only military, but also transport. As a result, they thereby gain superiority with regard to mobilization. What conclusion, then, follows from this? It is this, that our strategic plan—not an abstract plan, but one calculated for a concrete situation and concrete conditions—must envisage during the first period of the war not an offensive but the defensive. Its aim—to gain time for the unfolding of mobilization. Consequently we consciously leave it to our enemy to be the first to attack, not at all considering that he will thereby gain some "moral" preponderance. On the contrary, having space and numbers in our favor, we calmly and confidently fix the limit where our mobilization, secured by our flexible defense, shall have prepared a sufficient fist to enable us to go over to the counter-offensive.

The formulation of the French field statutes is obviously false. It speaks of the need of being the first to attack, evidently from the standpoint of the need of gaining tempo. It is incontestable that tempo is a very important thing in the bloody game of war. Chess players know how important tempo is on a field of 64 squares; but only an ardent young chess player believes that tempo will be won by him who begins to check-mate first. On the contrary, this is frequently the surest way of losing tempo. Should I be the first to pass over to attack while my attack is not sustained by adequate mobilization and if I find myself compelled to retreat and therewith to disrupt my own mobilization, then of course I shall have lost tempo and, perhaps, irretrievably so. On the contrary, if a preparatory retreat enters into my plan; if this plan is clearly understood by the senior commanding personnel, which is confident of what tomorrow will bring, and if this confidence is transmitted from the top to the ranks below, without running up against the prejudice of an alleged necessity of invariably being the first to attack, then I have all the chances of retrieving the tempo and of winning.

Point 14 which states that our most urgent task is to review our statutes, propositions and instructions from the standpoint of the experience of the civil war is absolutely correct. But this has been said by us three years ago and it has been sealed by the decision of the Party Congress; corresponding orders have been issued and agencies to review the statutes have been set up. Sad to say, the work is progressing rather slowly. It must be speeded up. But to inform us, under the guise of a new "military doctrine," that we must review the statutes when all the corresponding agencies for this review have long since been created, is truly to break needlessly into long open doors.

The practical conclusions at the end of the theses are by and large correct. But they do not at all flow from the premises and, in addition, they are inadequate; and they do not specify the central task—the securing of the army's stability and qualification through the education of the lower commanding personnel. What we need are individual platoon commanders! No matter what strategy may be imposed upon us by the march of events—whether it be maneuverist or positional, or a combination of both—the fundamental moment of combat operations remains the military section whose basic cell is the platoon, with the platoon commander at the head. This is the brick out of which, if it is properly baked, any edifice may be constructed.

What Is Old in the "Novelty"

Having read the theses of Comrade Frunze, I skimmed through Suvorov's "Science of Victory." The designation, "science" is of course incorrect; but Suvorov understood it in its most simplified form, that is, in the sense of that which must be assimilated. Precisely in this sense the soldier, when made to run the gauntlet, was admonished: "Here is science for you." Under Suvorov's dictation Lieutenant-General Prevost de Lumian wrote down seven laws of war. Here they are:

1. Act not otherwise than on the offensive.

2. When marching—speed is paramount; in attack—impetuosity, cold steel.

3. What we need is not methodism, but a correct military outlook.

4. All power to the commander-in-chief.

5. The enemy must be attacked and beaten in the field, that is, don't remain sitting in fortified regions but keep after the enemy.

6. Don't waste time on sieges. An open frontal assault is best of all.

7. Never divide forces for the sake of occupying points. If the enemy outflanks you, so much the better; the enemy is himself heading for defeat.

What is this if not a proletarian doctrine?! This is almost word for word a strategy that "flows from the class nature of the proletariat" and out of the civil war—only somewhat more succinctly and better stated! . . . Suvorov was of course in favor of the offensive. But he also said that we need not methodism but a correct military outlook . . . However, Suvorov, after all, led into battle a feudal army under the command of officer-nobles. It thus turns out that the principles of "the offensive doctrine of the proletariat" coincide not only with the field statutes of bourgeois-imperialist France, but also with the military "science" of the Suvorovist landlord-feudal Russia!

From this it does not at all follow that "the laws of war are eternal," as certain pedants say. Under discussion here are not at all laws in the scientific sense, but rather practical usages. Some of the simplest generalizations (as for example the advise—"attack, and do so impetuously") apply to all forms of struggle between living creatures. Rule of thumb, speed, aggressiveness are necessary not only during clashes between two organized and armed forces, but also during a fist fight between two little boys and even when a hunting dog chases a rabbit. But if the seven Suvorovist commandments are not eternal laws of war, then it is even less possible to pass them off as the most modern principles of proletarian strategy.

Is there a difference between the Red Army and the army of Suvorov? There is. An enormous one. Incalculable. In the one case you have a feudal army, kept in darkness. Here you have an army that is revolutionary, and whose consciousness is growing. The aims are diametrically opposite. We are undermining everything that Suvorov defended. But this difference involves not a military doctrine but a class political world-outlook. In this little book, in his aphorisms, Suvorov also expounds a social world-outlook. Lacking it, Suvorov would not have been an army leader. Suvorov's entire psychologic art consisted in extracting the most out of the instrument represented by a feudal soldier. In his social doctrine Suvorov rested on two poles: gauntlets and "God is with us." In their place we have the Communist program and the Soviet constitution.

Here we have made a certain step forward. And not a small one. On this score the Kharkov theses can hardly offer us something new. And besides, we do not feel any need of renovating our social world-outlook. So far as the questions of strategy are concerned, then here, as we see, the matter came down to this, that those who began by promising us a new proletarian doctrine, ended by copying out the rules of Suvorov, and made mistakes in copying.

II

SUMMARY SPEECH

Doctrine, Outlook, A Monistic Viewpoint

It is first of all necessary to occupy positions which are cleared by the opponent in his "maneuverist" retreat. This is the first thing . . .

Comrade Frunze admits that here and there his formulations are inexact, ambiguous, inconclusive. If it were a question of a draft of an article, then it is quite self-evident that such shortcomings would be perfectly natural. But when it is said that "you have no doctrine, whereas I do have a doctrine"—as Comrade Frunze poses (or used to pose?) the question—then you are dealing with something of an entirely different order. After all, at the Tenth Party Congress, Comrades Frunze and Gussev took me very severely to task for lacking interest in the

question of military doctrine, wherein, according to them, lay
the whole crux of the matter. At the time they thumped my
head lightly with a volume of Engels (without sufficient
grounds, but I leave this for another occasion). What, then,
to do? Engels came out as a theoretician of military affairs,
whereas we still continue to fight empirically. Well, show us
your "doctrine," Comrade Critics. But take care; it is possible
to fight with an oven-fork, for lack of a different weapon—but
it is impossible to write theory with an oven-fork; different
instruments are here needed. But after all, is anyone forcing
us to rush ahead with this question? There is no fire. True,
Comrade Frunze very delicately hints that, as you know, after
the Russo-Japanese war, there was an occasion when, by august
order, all discussions concerning military doctrine had to be
terminated and the study of statutes had to be undertaken. One
thus seems to arrive at a certain not very pleasant analogy:
Comrade Frunze proposes to take up the question of doctrine
while I "order" that unpleasant discussions be terminated and
a study of the statutes be undertaken.

But in reality, this juxtaposition is extremely arbitrary and
its barb can be turned against Comrade Frunze himself. For
what was the content of the task and aim of those Russian
officers who, after the Russo-Japanese war, began talking about
military doctrine? They represented the critical elements in the
army. They were dissatisfied with its structure and wanted
changes introduced. This was the progressive section of the officerdom, the very same ones who later united around Guchkov
and Miliukov and whom the Black Hundred men called "Young
Turks." Consequently, the banner of military doctrine was for
them the banner of criticism of the past and the program of
military reform. Insofar as it was possible they wanted to
Europeanize our army and in this connection sought support
even in the State Duma. They were ordered to shut up, not to
criticize, not to undermine autocratic Asiatism. And how do
matters stand with us? What is the content of Comrade Frunze's
military doctrine? It consists of an uncritical idealization of
the past. Our heralds of military doctrine seek to deduce from
the class nature of the proletariat, and to render eternal that
which characterized a certain period of the war. What did
Comrade Frunze accuse me of in his speech? Of denying to the
past the reverence it merits. He considers the idealization of
the past an indispensable element in the army's moral educa-

tion. But this was precisely the standpoint of those who inspired Czar Nicholas to issue his august order—to terminate discussion on doctrine, not to undermine the reverence of the past. But we say to you: Please stop threatening to annihilate the enemy by throwing hats at him; let us instead learn the ABC of military affairs from the enemy. This is where the basic disagreement lies and this is what Comrade Frunze refuses to assimilate.

By way of compensation Comrade Minin has enriched us with a new term: if we reject a unified military doctrine, if Comrade Frunze is ready to reject also a military world-outlook, then Comrade Minin offers us a "monistic viewpoint" on military affairs. This has a proud ring: a monistic viewpoint is hardly inferior to your term, doctrine. But what do you mean to say by it? That a unity of views, usages and methods is needed within the framework of an army? Why, of course. It is needless to waste eloquence in order to prove that an army is incompatible with such an order—or disorder—as one pulling one way while the other pulls the other way. Have we got agreement then? A unity of methods is necessary; let us call this unity "doctrine"—and that is all! Such a proposal was virtually made by Comrade Kashirin: it is necessary for the state to define its views on war in a single doctrine. Is the whole dispute then seemingly over words? Oh, no. The gist of the dispute goes much deeper—it lies in the confusion of concepts. What do you mean in the last analysis by military doctrine? Do you mean the answer to the question of *what* we are fighting for? or the answer to the question *how to fight?* or finally the answer to both these questions together? *(Kashirin interjects: "Both questions.")* That's it precisely—you need a military doctrine in the sense of some sort of answer to "the meaning and aims of the war." Here you are wholly captives of the bourgeois state. Inasmuch as the bourgeois state conducted and continues to conduct wars for the sake of plunder and enslavement, it was compelled to motivate the genuine aims of the war by a special and ostentatious "national military doctrine." The task of this doctrine is to deceive the popular masses, hypnotize them, render them blind.

The English doctrine is: the civilizing role of the Anglo-Saxons throughout the entire world and especially in the colonies; the highest interests of culture demand that Great Britain rule the seas; hence—the English fleet must be stronger

than the next two most powerful fleets. Behind this military
doctrine lurk the class interests of the bourgeoisie. Is there any
need for us to create a special doctrine in order to explain
why and for what we have to wage wars! Not the slightest.
We have the Communist program; we have the Soviet constitution; we have the land law—there's our answer. What more
do you need? Is there another country with an answer which
in any way approximates in power the answer given by our
revolution? Our revolution has destroyed the ruling, possessing classes; it has handed the power over to the toilers and
said: Defend this power, defend yourselves—here are your war
aims.

Captives of Bourgeois Ideology

You are demanding that the Red Army pose itself a goal in
the shape of some kind of doctrine—meanwhile the revolution
in order to serve its own needs has created out of us an army
and has ordered us to study military affairs as they must be
studied; to fight as it is necessary to fight. And we did fight
for more than three years. But when things eased up a little
we began pondering over a profound question: where can we
find a doctrine that would explain to us *for what* we shall go
on fighting? This is ludicrous pedantry! There is a second
question: *how to fight.* Here we are told that it is necessary
to have a unity of methods. Yes, of course! Why else did we
conduct the struggle against guerrillaism, parochialism and
superficial "independent" notions? Why else did we create a
centralized apparatus headed by the Revolutionary Military
Council of the Republic? For the sake of what did we write
statutes and regulations and establish tribunals? On number-
less occasions it was necessary for us (including myself) to
explain and to prove that a unity of poor methods is superior
to a diversity even of the best methods. I had to prove this in
the struggle against guerrillaism in Tsaritsin, too, which is the
home town of Comrade Minin who now *objects to having one
pull one way while another pulls another way.* In those days
some of the present adherents of military doctrine used to
declare that they would carry out good orders at the front but
would refuse to carry out orders they deemed incorrect. In
those days it was necessary to deal severely with self-opinionated
commanders of divisions and of brigades who grew up in the
atmosphere of guerrillaism and who refused to grasp the mean-

ing of the unity of organization and the unity of methods. All
our efforts throughout the existence of the Red Army came
down precisely to guaranteeing the maximum planfulness, the
highest unity, the closest harmony. After all, this was the end
which was served, and continues to be served by all our statutes, formations, regulations, orders, circulars, instructions, in-
spection commissions and tribunals. And even at the present
time a considerable part of the interrelations between the
Supreme Military Council of the Republic and the military
districts and fronts consists in the *struggle against deviations
from those formations and norms* which have been established
by the center. Naturally, our statutes and formations are not
absolute. We shall review them in the light of our experience.
In reviewing and improving our methods we thereby defend
their unity. By transferring the question to the plane of elementary discussions concerning the usefulness of the unity of
methods, you are actually throwing us back three years—back
to the period of our struggle against guerrillaism and parochialism; and this is being passed off by you as some kind of new
military doctrine.

Comrade Kuzmin has dealt with the question of offensive
and defensive warfare. And it turns out that there are no
difficulties here at all. Comrade Kuzmin is able to dispel all
the trouble by a mere wave of the hand. Trotsky, you see,
argues against offensive revolutionary war and is in favor of
the defensive. But now, I, Kuzmin, will say to the Red Army
soldiers, to the workers and peasants: "Russia is today a beleaguered fortress; you are her garrison, but tomorrow it will
perhaps be necessary for you to go from the stronghold into the
field in order to break the blockade!" And that's all; it is as
simple as that. But after all, Comrades, this is not a serious
political attitude to the question, but completely that of a writer
of feuilletons. The issue is presented as if it were enough to
find a suitable simile, a military image, in order to dispel all
difficulties with a gesture . . . No, this is not the crux of the
matter at all. It is only necessary to clearly separate the
political question from the strategic. Politically we are firmly
maintaining a defensive position. We do not want war, and
the entire population of our country must know and understand
this. We are taking all possible measures in order to avoid
war. We proclaim our readiness under certain conditions to
pay the Czarist debts.

Attack and Defense

I recall that one comrade said to me, "Why do you say openly that we are ready to acknowledge *Czarist debts?*" This comrade seemed to be embarrassed by our being forced to agree to such a concession, and he sought to proffer it to the workers and peasants in a masked form. This is a crude blunder. One must speak clearly, simply and frankly. And in the long run this will be only to our advantage. We say to the workers and peasants: "Payment of Czarist debts is being demanded of us. The Czar took money from the stock-market in order to strangle you, workers and peasants; and now it is demanded of you, workers and peasants, that you pay for being strangled by the Czar. And we, the Soviet power, are prepared, under certain conditions, to agree even to the payment of these abysmally dishonest, bloody debts. Why? Because we wish to spare our country the ravages of a new war."

In this way we clarify to the peasants the peaceful and defensive character of our policy. Armed bands have been thrown against us. We destroyed these bands, but did not go over to the offensive. We have truly revealed and continue to reveal an incredible forbearance. Why? Because we want to secure peace to the people. And this is now the foundation of our political-educational work in the army and in the country. And what if peace is denied us? What if we are compelled to wage war? In that case the most backward peasant will understand that the blame falls wholly on our enemies, that there is no other way out; he will then take his hunting spear and march into battle. Then, too, it will be possible for us to unfold an offensive war in the strategic sense. Then the Red Army soldier, the worker and the peasant will say: "Our entire policy was directed toward defense and toward peaceful relations. But if these neighbors, these governments refuse us peace, despite all our efforts, then for the sake of defense, nothing remains for us except to beat them down . . ."

Such will be the extreme conclusion of the entire country in the event that our defensive and peace-loving policy is disrupted by our enemies. Herein is the essence of the question. He who understands this will find the correct line for political work in the army. But allegories about a beleaguered fortress will avail little here. It is only a metaphor, an image for a leading editorial or a feuilleton. A Samara moujik on reading this, or hearing someone else read it aloud, will scratch the back of his head and say, "A clever writer is Comrade Kuzmin; he writes fine." But for the sake of this metaphor, I assure you, he will not go to fight.

Comrade Voroshilov cited here my words to the effect that under certain conditions the road from Petrograd to Helsingfors may prove to be shorter than the road from Helsingfors to Petrograd. Yes, it is true I said this; and under certain conditions I am ready to repeat it again. But, after all, this is precisely what I have just been explaining. This does not at all mean that we are actually preparing to attack any of the neighboring countries. This is excellently understood by you. True enough, in the frontier regions where our warriors have had the occasion to observe very closely the banditism of Polish, or Roumanian, or Finnish origin, the moods in favor of a blow across the frontiers are at times quite strong among our troops. "Let's have war!" These words are often to be heard there, especially among cavalry men. Our military students are likewise not averse to verifying in practice what they are learning in theory. Moreover, throughout our whole army there fortunately prevails the mood of readiness for battle.

But, after all, this does not exhaust the question. War is a big, serious and lengthy business. It presupposes new mobilizations of several draft-ages, the mobilization of horses, the redoubling of land-carried levies, etc., etc. It is absolutely self-evident that we cannot start a war with the propaganda of the correct, abstract idea that the interests of all toilers in the world are identical, and so forth and so on. This idea is correct and it must occupy the most prominent place in our propaganda, above all within our own party. But there is an enormous difference between the propaganda of the idea of the world revolution and the political preparation of the toiling masses of the country for military events which may possibly occur in the immediate future. This difference is the difference between propaganda and agitation, between a theoretical forecast and current policy. The more clearly, persistently and concretely, the more unquestionably we are able to show and clarify to the entire population of the country the genuinely defensive character of our international policy, all the better prepared will be the entire population to provide the forces and resources for an offensive strategy on a broad scale, in the event that war is nevertheless foisted upon us. Comrade Frunze

does not argue against this. On the contrary, he has even declared that it would be the most stupid project to talk today
of an offensive war on our part. This is correct. But you
have only to read some of the most recent articles of Comrade
Frunze's closest co-thinkers to find stated there that up to now
we have been "sitting" on the defensive, but that now we are
preparing for the offensive. It is very good that Comrade
Frunze has definitively and even sharply differentiated himself
from this false *political* point of view which cannot bring us
anything except difficulties, confusion and harm.

But isn't it impermissible to renounce the idea of political
offensive in general? Why, of course! We are not in the
slightest preparing to renounce the world proletarian revolution
and the victory over the bourgeoisie on a world scale. We
would be traitors and betrayers, like the gentlemen of the
Second and Two-and-a-Half Internationals if we renounced the
revolutionary offensive. But, after all, the reciprocal relations
between the preparatory, defensive work and the offensive have
been sufficiently fully and clearly elaborated on an international
political scale at the Third World Congress of the Communist
International. The adherents of the doctrine of the offensive
were present there, too. They also said: "The offensive corresponds to the revolutionary nature of the working class or
the character of the present revolutionary epoch." And when
we set siege to them and set about to curb them, these "leftists"
began to cry out: "Ah! So you renounce the offensive?" We
renounce nothing at all, dear Comrades. But all in good time.
Without the offensive, victory is impossible; but only a simpleton believes that the entire political tactic comes down to the
slogan: "Rush Ahead!"

In the Grip of "Sad Necessity"

The idea of a revolutionary offensive can be tied up only
with the idea of an international proletarian offensive. But
is this the current slogan of the Comintern? No, we have
advanced and we are defending the idea of a working class
united front, of joint actions even with the parties of the Second
International who do not want the revolution—and this, on the
basis of defending today's vital interests of the proletariat which
are being threatened from all sides by the aggressive bourgeoisie.
Our task is to conquer the masses. How is it, Comrades, that
you have overlooked this tactic, failed to assimilate its meaning,

failed to clarify its connection with the new economic policy within our country? It is quite self-evident that at present it is a question of major preparatory work, at the given moment defensive in character and of the broadest mass sweep. From this work there will inevitably grow at a certain stage the mass offensive led by the Communists; but today this is not the task. You ought to bring our military propaganda in harmony with the general course of the policies of the world working class. It is stupid to talk to the Red Army about a revolutionary offensive at a time when we are summoning the European Communist parties to make careful preparations on an ever wider mass basis. When a change occurs in the world situation, the slogan of our educational work will change with it.

This is how matters stand today with regard to the question of the offensive in a political sense. But there still remains the strategic and tactical side of the question. And here, after all of Comrade Frunze's explanations, I remain wholly on the standpoint that the formula of the French field staff is wrong: it suffers from formalism with respect to the offensive. The idea of the offensive is expressed far more correctly in our own field statutes. "The best method of gaining a goal is to act aggressively." Nothing is said here to the effect that he who attacks *first* allegedly "reveals a much stronger will." The task of war is to annihilate the enemy. Annihilation is impossible without the offensive. The stronger will is revealed by him who creates the most favorable conditions for the offensive and utilizes them to the very end. But this does not at all mean that in order to reveal his will one must attack first. This is nonsense. If the material conditions of mobilization militate against it, then I would be a hopeless formalist and a dunderhead to build my plan on the notion that I must be the first to take the offensive. No, I shall reveal the superiority of my will by creating favorable conditions for my offensive—as the second one; by wresting the initiative at a certain limit fixed in advance; and by gaining victory even though I am the second to attack. (*Frunze interjects:* "This is *less advantageous.*")

This may be less advantageous in relation to an *abstract* country with different railways and a different apparatus of mobilization from ours; but, after all, we are engaged not in solving a geometrical problem but in outlining a concrete plan of action, depending upon the material and spiritual conditions of our country and its reciprocal relations with other countries.

On the one hand, Comrade Frunze in every way emphasizes that
we will enter into combat, equipped with a lower technology
than our enemies; and he seemingly even introduces this lower
technology into our military "doctrine." We must of course do
everything in our power to raise our technology to the level of
our enemies. But it is quite understandable that they will have
the preponderance with respect to, say, aviation. Comrade
Frunze takes this into account, emphasizes this in every way, and
as a means of counteracting it recommends, for example, that
our troops be trained for night operations. Why then does he
forget about the condition of transport which happens to be
under the existing conditions the most important part of military
technology? It is impermissible to forget about mobilization,
concentration and deployment of forces. Serious strategy must
take precisely this as its starting point. The necessity to attack
is beyond dispute. This is stated not only by our statutes but
also by the old Czarist ones, and almost in the same words.
We have heard this from the lips of Suvorov. How is it possible to vanquish the enemy except by dealing him a blow over
the head? And for this it is necessary to attack him, to spring
upon him. This was known to army leaders in biblical times.
But it is your desire to communicate something new to us, you
talk to us about a proletarian strategy flowing from the revolutionary nature of the proletariat. You are apparently not satisfied with the formula in our field statutes. You create a
formula of your own which—surprise of surprises!—happens to
be borrowed from the French field statutes. But this allegedly
new formula is false and it obviously does not correspond to
our conditions. If we hammer into the minds of our commanding personnel that revolutionary nature and "strong will" demand that you be the *first* to attack then the very first period
of our operations in the West can lead our commanding staff
astray because conditions may impose upon us, and in all
likelihood will impose upon us an initial period of flexible
defense and maneuverist retreat. *(Frunze: "Sad necessity!")*
Yes, Comrade Frunze, war in its entirety is a sad necessity.

Within the framework of this sad necessity it is necessary
to build one's plans, taking into account other "sad necessities,"
if they are of cardinal importance. And the condition of
transport, in the broadest sense of the word, is one of the
cardinal conditions of war. In consequence, the nature of our
country, its distances, the distribution of its population, its rail-

ways, its macadam and country roads make it quite probable that the starting point of our offensive will be a line at a considerable remove from our state frontiers. If our commanding staff grasps the inner logic of such a strategic plan which begins with screening operations, defense and even retreat in order to consolidate the troops along a border-line fixed in advance and then to pass over to a decisive offensive without which victory is, naturally, impossible; if our commanding staff becomes imbued with this genuinely maneuverist idea and not with a formalistic attitude toward the offensive, then they will not be disorganized nor led astray nor lose their heads but transmit their calm assurance to the entire army.

Our Agitation as a "Kind of Weapon"

Revolutionary agitation as a new kind of weapon introduced by us has been adduced here in support of the contention that we have our own "military doctrine." But this, too, is false. We are deceiving ourselves here as well. As a matter of fact, propaganda in bourgeois armies is arranged on a far broader scale, much more richly and diversifiedly than is the case with us. During the first two years of the [last] war I lived in France and had the opportunity to observe there the mechanics of imperialist agitation. How can we possibly compete with it in the face of our poverty of forces and resources? Our newspapers are tiny, the paper is poor, the print extremely illegible, and, most important, their circulation, insignificant. Whereas in France such an obscenely mendacious, brazen bourgeois newspaper as *Petit Parisien* used to come out during the war in printings of almost three million copies. The circulation of several other imperialist newspapers was over a million. Every soldier received a newspaper, not infrequently two. Here was poetry and prose; feuilletons and cartoons. And the newspapers played in all the colors of the rainbow: Monarchist as well as Republican as well as Socialist. And they all kept hammering at a single point: war to the end. Right there was the Catholic priest walking through the trenches and functioning as a very skillful agitator, patting the soldier on the back and telling him: Only two good things are left in this world— wine and the Lord Almighty. A Socialist deputy arriving at the front would talk about the struggle for freedom, equality, and so forth and so on. Also there was the theater, the ballet, the chorus girls. And all of it, first-rate. And all of this hammering away at a single point. What a monstrous machine of deception, hypnosis, catalepsy and degeneration! Wherein then does our strength lie? In the *Communist* program. In the *revolutionary* idea. When our enemies talk about the monstrous power of our propaganda, it applies not to our organization or technique of propaganda in the army but to the *inner* strength of our revolutionary program which expresses the genuine interests of the toiling masses and therefore touches them to the quick. It was not we who invented politics. It was not we who invented agitation and propaganda. In this respect likewise, our enemies are materially and organizationally stronger than we are, just as Czarism was far stronger than our party when it was underground and had to function through circulars and proclamations. But the whole gist of the matter is this, that with all its apparatus and all its technology, the bourgeoisie cannot maintain its hold on the masses. But we are conquering and shall conquer them throughout the whole world. There is no need therefore of discovering a new kind of weapon, entering into the military doctrine of the proletariat. For the Communist program was discovered before the Red Army arose, and the Red Army itself is only a weapon for assuring the possibility of realizing the Communist program in life.

Fewer Sweeping Generalizations

The connection between strategic and tactical methods and the class nature of the proletariat is not at all so intimate, so unconditional and so immediate as many comrades here have expounded. On the basis of my personal and admittedly modest knowledge of the history of military affairs I would undertake to prove that the Red Army from the outset of its existence has passed through those stages which have marked the evolution of modern European armies, say, since the seventeenth century. Naturally, the transition from one stage to the next was accomplished very rapidly, as if in a telescoped form. A child in its mother's womb, as it develops from the embryo, repeats the stages of the evolution of the human species in its fundamental features. Something similar, I repeat, may be observed in the development of the Red Army. It did not at all begin with maneuverability. Its initial combat attempts give us a picture of an angular, crude positionalism of a cordon type. Its organizational and strategical methods underwent change in the process of the struggle, under the blows of the enemy. In this way there unfolded the maneuverability of the last period of the civil war. But this is not the last word of the Red Army's strategy. Into this amorphous, chaotic maneuverability we must introduce the elements of stability: firm, flexible cadres. Will this more highly qualified army arrive at positional methods? This depends on the conditions of future wars; this depends on where these wars will take place, the size of the masses that will be simultaneously drawn into war operations, and the territories on which the latter will unfold.

Comrade Budenny found the explanation for the positionalism of the imperialist war in an absence of great initiative, in the indecisiveness of the leadership. "A genius army leader was lacking! . . ." In my opinion this explanation is erroneous. The crux of the matter lies in this, that the imperialist war was a war not of armies but of nations, and therewith, the richest nations, huge in numbers and with huge material resources. It was a war to the death. To every blow the opposing side found an answer; every hole was plugged up. The front was constantly reinforced by both sides; artillery, munitions, men were piled up both here and there. The task thus went beyond the bounds of strategy. The war became transformed into the most profound measuring of reciprocal forces in every direction. Neither aviation, submarines, tanks nor cavalry could in and by themselves produce the decisive result; they served only as the means of gradually exhausting the enemy's forces and of constantly verifying the enemy's condition: is he still able to maintain himself or is he ready to collapse? This was in the full sense of the word a war of attrition in which strategy is not of decisive but of auxiliary significance. It is quite indisputable that a repetition of such a war in the immediate future is impossible. But just as impossible is a repetition on European territory of the methods and usages of our civil war: the conditions and environment are much too different there. Instead of making sweeping generalizations we ought to start thinking more definitely about concrete conditions.

"Unified Doctrine" in a Future Civil War

For the sake of illustration let us take England and let us try to imagine what will be, or more correctly, may be the character of a civil war in the British Isles. Naturally, we cannot prophesy. Naturally, events may unfold in an altogether different way, but it is nevertheless profitable to try to imagine the march of revolutionary events under the peculiar conditions of a highly developed capitalist country in an insular position.

The proletariat constitutes the overwhelming majority of the population in England. It has many conservative tendencies. It is hard to budge. But in return, once it starts moving and after it overcomes the first organized opposition of internal enemies its ascendancy on the islands will prove to be overwhelming owing to its overwhelming numbers. Does this mean that the bourgeoisie of Great Britain will not make the attempt with the assistance of Australia, Canada, the United States and others to overthrow the English proletariat? Of course it will. For this, it will attempt to retain the navy in its hands. The bourgeoisie will require the navy not only to institute a famine blockade but also for purposes of invasion raids. The French bourgeoisie will not refuse black regiments. The same fleet that now serves for the defense of the British Isles and for keeping them supplied uninterruptedly with necessities will become the instrument of attack upon these islands. Proletarian Great Britain will thus turn out to be a beleaguered naval fortress. There is no way of retreat from it, unless into the ocean. And we have presupposed that the ocean will remain in enemy hands. The civil war will consequently assume the character of the defense of an island against warships and invasion raids. I repeat this is no prophecy: events may unfold in a different way. But who will be so bold as to insist that the scheme of civil war outlined by me is impossible? It is quite possible and even probable. It would be a good thing for our strategists to ponder over this. They would then become completely convinced how unfounded it is to deduce maneuverability from the revolutionary nature of the proletariat. For all anyone knows, the English proletariat may find itself compelled to cover the shores of its islands with trenches, deep ribbons of barbed wire defences and positional artillery.

Models of civil war approximating our recent past, we ought to seek not in Europe of the future but in the past of the United States. It is unquestionable that the civil war in the United States in the 'sixties of the last century discloses many features in common with our civil war. Why? Because there, too, you had enormous spaces, a sparse population, inadequate means of communication. Cavalry raids played an enormous role there, too. It is a remarkable thing that there the initiative likewise came from the "Whites," that is from the Southern slaveowners who waged war against the bourgeois and petty-bourgeois democrats of the North. The Southerners possessed prairies, plantations, prairie pasture lands, good horses and were accustomed to riding horseback. The initial raids, thousands of versts in depth, were executed by them. Following their example, the Northerners created their own cavalry. The war was of a diffused, maneuverist character and terminated in the victory of the Northerners who defended the progressive tendencies of economic development against the Southern plantation slaveowners.

En Route to Proletarian Strategy

Comrade Tukhachevsky expressed himself in basic agreement with me, but made certain reservations the meaning of which is not clear to me. "That Comrade Trotsky," says Tukhachevsky, "keeps pulling back by the coattails is a useful thing." But this is useful, insofar as I am able to gather, only up to a certain point; for the very urge to create something new, in the sense of proletarian strategy and tactics, seems to Tukhachevsky an urge that is fruitful and progressive. Comrade Frunze, marching along the same line but going much further, cites Engels who wrote in the 'fifties that the conquest of power by the proletariat and the evolution of socialist society will create the premises for a new strategy. I also have no doubts that if a country with developed socialist economy were compelled to wage war against a bourgeois country (as Engels visualized the situation in his mind) the picture of the socialist country's strategy would be one that is entirely different. But this provides no grounds whatever for attempts to suck out of one's thumb a "proletarian strategy" for the USSR today. A new strategic word grows out of the urge to improve and fructify the practice of war and not at all out of the mere urge to say "something new." This is similar to someone who values original people setting himself the task of becoming original himself—naturally, he would not attain anything except the most wretched monkeyshines. By developing socialist economy, by raising the cultural level and fusing the ranks of the toiling masses, by raising the qualifications of the Red Army, by improving its technique and cadres, we shall undoubtedly enrich military affairs with new usages and new

methods—precisely because our entire country will grow and
evolve on new foundations. But to set oneself the task of
deducing through speculation a new strategy from the revolutionary nature of the proletariat is to occupy oneself with
patching up dubious propositions of the French field statutes
and inevitably to lose one's bearings.

Forward to the Accumulation of Culture!

In conclusion I want to dwell on the question of the individual platoon commander. Everybody of course recognizes
the importance and meaning of a platoon commander but not
everybody is willing to see in him the central point of our
military program in the period immediately ahead. Some
comrades even express themselves somewhat condescendingly on
this question: "Of course, who would deny . . . Yes, of course
. . . Yes, obviously . . . But the world is large enough to
contain more people than just the platoon commander . . ."
And so forth and so on. The remarks of our charming Comrade Muralov smacked a little of this spirit. He said: "Naturally, it is necessary to grease boots, sew on buttons and educate
good platoon commanders, but this is far from everything."
For some unknown reason, the platoon commander is here
lumped together with buttons and boots. In vain! Buttons,
boots and the like pertain to those "trifles" which in their
totality are of enormous importance. But the platoon commander is in no case a trifle. No, this is the most important
lever of our military mechanics.

But in passing, allow me first to say a few words on buttons,
boots, the war on lice, etc. Comrade Minin accused me of
falling into cultural-uplift spirit. It is too bad he failed at
the same time to indict also Comrade Lenin for his report to the
Party Congress, for Lenin's main idea was that we lack sufficient culture for our work of construction and that we must
persistently, stubbornly and systematically accumulate this culture, raising it through education and self-education. The term
"cultural-uplifter" does not apply here because we used it
against and even branded with it those curmudgeons who under
the rule of Czarism and of the bourgeoisie hoped to regenerate
the country through trifling and petty educational, cooperative,
sanitary and similar measures. We counterposed to this the
program of the revolution and the conquest of power by the
working class. But today this has been achieved, power has

been conquered by the working class: this means that political conditions have been created for cultural work on a scale hitherto unprecedented in history. This cultural work comprises exclusively of details and trifles. The victorious revolution gives us the opportunity to draw into cultural work the thickest nethermost layers of the people. This is now the main task. It is necessary to teach how to read and write; it is necessary to teach precision and thriftiness—and do all this on the basis of the experience of our state and economic construction, day by day, hour by hour. And exactly the same thing in the army.

Today's Military Slogan

But the *platoon commander*—that is still a special item. This is by no means a trifle; this is the commander, the leader, the head of the basic military group—the platoon. It is impossible to build an edifice with loose sand. One must have good building material, one must have a good platoon, and this means—a good, reliable, class-conscious, confident platoon commander.

"But," some object, "aren't you forgetting about the senior commanding staff?" No, I am not forgetting it and it is precisely to the senior commanding staff that I set the task of educating the platoon commander. There can be no better school for a commander of a regiment, or a brigade, or a division than the work of educating the platoon commander. Our post-graduate courses, our academies and academical courses are very important and useful, but the best training of all is received by a teacher in teaching his pupils; best trained of all will be the commander of a regiment, the commander of a brigade and the commander of a division who centers his attention in the immediate future on the training and education of platoon commanders. For this cannot be done without clarifying more and more in one's mind all the questions of Red Army organization and tactics without a single exception.

One must think out clearly for himself all the questions, think them out to the end, without any self-deception in order to be able clearly and precisely to tell the platoon commander what he must be and what is demanded of him. The platoon commander—this is now the central task. General phrases about educating the commanding personnel in the spirit of maneuverability offer very little in essence, distracting attention away from the most important task of the present period. There was a time when it was necessary to smash our primitive immobility and our cordonism; there was a time when the slogan of maneuverability was salutary; at that time the cry: "Worker on Horseback!" expressed a basic need. Of course, at that time not only the cavalry but also the infantry, the artillery, etc., were of importance. However, had we not at that time created the Red cavalry we would have probably perished. For this reason the summons "Worker on Horseback!" summed up the central, basic need of that particular period in our Army's development. The new epoch advances to the fore a new task: to set in order the basic cell of the army—the platoon; to sum up our military experience for the individual platoon com-mander, raise his knowledge, his self-esteem. Everything now centers about this point. It is necessary to understand this, and to get firmly to work.

4. MARXISM AND

MILITARY KNOWLEDGE

May 8, 1922

I

OPENING REMARKS

Permit me to declare open the 51st session of the Military Scientific Society.

The subject of today's discussion will be: The place of military knowledge and military skill in the system of human knowledge as a whole. Let me confess at the outset that the responsibility for initiating this discussion falls largely upon me. Not that I consider this complex, abstract, theoretical-epistemological and philosophical question—in the best and worst meaning of these words—to be the most current and unpostponable of our military studies. But it does seem to me that such questions are forced upon us by the entire course of ideological development, and a certain theoretical-ideological controversy among our army tops.

In one of our publications, closely associated with your Society, I happened to read two articles,* one of which presented the argument that military science cannot be built and the methods of Marxism cannot be applied to its tasks, inasmuch as military science pertains to the order of natural sciences. Accompanying this article was a polemical and critical reply, apparently reflecting the views of the editors.

---

* *Red Army*, No. 12, March 1922.—*Ed.*

In contrast, this reply was an attempt to prove that the methods of Marxism are *universal* scientific methods—and therefore retain their validity in the field of military science. Let me again confess that both these viewpoints seemed incorrect to me: Military science does not belong among natural sciences, because it is neither "natural" nor a "science." Our discussion today may perhaps bring us closer to clarification on this question.

But even if one grants that "military science" is a *science*, it is nevertheless impossible to grant that it can be built with the methods of Marxism; because historical materialism isn't at all a universal method for all sciences. This is the greatest possible misconception which, it seems to me, can lead to the most harmful consequences. It is possible to devote an entire lifetime to military affairs very successfully, without ever devoting any thought to theoretical-epistemological methods in military matters—just as I am able to take daily readings of my watch without knowing anything about its internal workings, its interplay of wheels and levers. If I know about the numbers and the hands, then I can't go wrong. But if, not satisfied with the movement of the hands on the dial, I want to talk about the construction of the watch, then I must really be acquainted with it; there can be no room for independent thinking here.

A Correct Attitude to Philosophy

In the course of a previous discussion (on unified military doctrine) I adduced one of the traits of George V. Plekhanov, the first crusader for Marxism on Russian soil, a man of broad vision and high gifts. Whenever Plekhanov observed that questions of philosophic materialism and historical materialism were being opposed to one another, or on the contrary lumped together, he hotly protested. Philosophic materialism is a theory imbedded in the foundation of *natural* sciences; while historical materialism explains the history of *human society*. Historical materialism is a method that explains not the structure of the entire universe, but a rigidly *delimited* group of phenomena; a method that analyzes the development of historical man. Philosophic materialism explains the movement of the universe as matter in the process of change and transformation; and it extends its explanation to include the "highest" manifestation of the spirit. It is difficult, if not impossible, to be a Marxist in politics and remain ignorant of historical materialism. It is quite possible to be a Marxist in politics and not know about philosophic materialism; such instances can be adduced to any number. . .

And whenever any Marxist (in our old terminology, "social democrat") used to stray into the domain of philosophy and began muddling there, the deceased Plekhanov would go after him without mercy. How many times was he told,

> "But, after all, George Valentinovich, this happens to be a very young man who hasn't had the time for questions of philosophy; he was busy fighting in the underground."

But Plekhanov would with reason reply:

> "If he doesn't know, then let him keep quiet. Nobody is forcing him to open his mouth. . . . There is nothing said in our program about a social democrat's having to have all his four feet shod with philosophic materialism. As a party member, you must be active; you must be a courageous fighter for the workers' cause; but once you do invade the field of philosophy, beware of muddling. . . ."

And Plekhanov would rise to his full height and reach for his superb polemical whip. Anyone reviewing the history of our party could still find discernible to this very day the marks left by this whip on many ribs.

My premise is that we should follow in the excellent tradition of the deceased Plekhanov in the field of applying philosophy to military affairs. We are not at all obliged to occupy ourselves with questions which are known as "gnosiological," "theoretical-epistemological," philosophical; but once we do take them up, then it is impermissible to muddle, and to go wandering with wrong instruments into an entirely different field in the attempt to apply the method of Marxism directly to military affairs, in the proper meaning of this word (not military politics).

It is the greatest misconception to try to build in the special sphere of military matters by means of the method of Marxism; no less a misconception is the attempt to include military matters in the list of natural sciences. Unless I am mistaken, the proponents of both these tendencies are ready to take the floor today; in all likelihood, they will be able to expound their views better than I can. After they have spoken, we shall take up the discussion.

Lessons of the Previous Discussion

I don't think, Comrades, that we shall arrive today at any binding decisions on this question. But if we do succeed in introducing some clarity into the issue, and if we draw the conclusion that caution must be exercised in applying Marxism directly in special creative spheres, then this alone would be a major conquest. In our discussion over "military doctrine," which has a certain bearing on today's question, we kept, as you all know, circling around and muddling to our heart's content; and I don't think we were greatly enriched thereby—unless in the negative sense only: all became convinced that nothing really significant came of it. We undertook to build a "unified military doctrine" on a "proletarian, Marxist" foundation, and after debating the matter, we retraced our steps and decided to review our statutes on the basis of our past experience. And we are reviewing them—slowly, limping along the roads and also into the pits, since our roads are rough country roads and there is no lack of ravines.

But I firmly hope that real benefits shall accrue from this review of our statutes. We will not think up a new military doctrine by means of special commissions, but, by way of compensation, we shall get rid of a lot of rubbish and set down more precise formulations in some things. So far as our today's session is concerned, the benefits of discussing the broad question of the relation between military affairs and Marxism will be rather those, so to speak, of mental hygiene: There will be less confusion. And in practical terms our task is: Let us learn to speak more simply about the cavalry; let us not clutter up our discussion of aviation with ostentatious Marxist terminology, high-sounding terms, pompous problems which turn out, one and all, to be hollow shells without kernel or content. . . .

This concludes, Comrades, the introductory remarks which I have taken the liberty to make. For the sake of the audience comprised of comrades, acclimated to questions of philosophy in varying degrees, I must very urgently request all reporters and those who take part in the discussion to express themselves in the most concrete terms as precisely, simply and lucidly as possible. I believe that I come quite close to the truth in saying that not everybody here has studied philosophy, so to speak, from beginning to end; and assuredly, some of us have not even read the most elementary books on philosophy. I believe that such a presentation, that is, one designed for an audience not expert in philosophy, will also have the added advantage of helping us examine the content of each reporter's kit bag. For philosophic terminology is an artifice akin to make-up. . . . The make-up may be terribly imposing but underneath it there is nothing at all. Yet, as I have had occasion to note from many articles in our military publications, this occultism for the augurs, this occult procedure for the initiated, these medieval traditions and practices are retained among us. And so, I ask you to expound your ideas as simply as possible.

With your permission, Comrades, we shall proceed with the discussion.

II

TROTSKY'S SUMMARY SPEECH

The speakers' list has been concluded. Allow me in summation to say a few words in defense of an art which, in my opinion, has been slighted here, slighted at the expense of military science, which several comrades have in their turn defended against our slurs, in my opinion imaginary.

Comrade Ogorodnikov, the last speaker, and a few others before him directed their attack especially at Comrade Svechin, against whom I, too, have had occasion to polemicize. They are indignant: How could a guild member of military science suddenly renounce himself, uncrown military knowledge and declare that there can be no talk of science here?

In a roundabout way Comrade Polonsky also touched upon this question. Let us get oriented, he says:

"Knowledge is either scientific or non-scientific. If military matters are scientific then we are dealing with a science; if they are unscientific, then...they are worth a groat."

Comrade Polonsky compared an army leader to a surgeon. Not a bad comparison! A surgeon performs an operation. This is an action which requires certain habits, a certain art; but for a student, watching the operation, says Comrade Polonsky, it is a science. But that isn't so at all. For the student, too, the operation is not a science but a schooling. If a painter makes a sketch, then this is art; others sit around and copy. What would you say this is for them, for the students? Is it a science? No. It is a schooling, which is not quite a science. This is the way in which "science" was understood in Suvorov's days when the soldiers were made to run the gantlet. This was even known as the "science of victory."

One of the speakers said that it was impermissible to compare military affairs to art. Art, if you please, has esthetic criteria. And what about the practical arts? What about the art of building bridges, the art of building houses, the art of canalizing? A practical art, let us not forget, also has a scientific basis. In the last analysis all sciences have, of course, grown out of practice, out of the crafts and other varied activities; later on, however, they freed themselves from this direct, "coarse" association, while nevertheless preserving their historical, utilitarian significance. In making chemical experiments or following the crossing of species in a laboratory, a scientist may be pursuing an immediate practical aim, but he also may not be. On the other hand, even a purely theoretical conclusion serves in the last analysis to enrich practice.

Art and Science

An art may base itself on a multiplicity of sciences. One man works in science for the sake of science, "selflessly" as the saying goes; another operates with scientific conclusions only for practical goals; a third, guided by creative instinct, catches up intuitively what he requires for practice. Comrade Snessarev hit the nub better than the others when he proposed to apply the term "science-ized art" to military affairs. A dozen other terms can of course be devised, nor do I propose to make Snessarev's term obligatory but, in my opinion, the author of this term showed himself freest from guild prejudices when he said, "Even the denomination of a craft does not scare me; all the less so will I shy away from the denomination of an art."

Many comrades approached the question under discussion from an "aristocratic" standpoint, from the standpoint of commanders—military leaders of today or tomorrow. But if we take military affairs as a whole, then the fact remains that every soldier must know his maneuver. That maneuver which a *rank-and-file* infantry soldier knows or has to know—is this a science or no? It is said about a *commander* that he must know geography and history—it would not be amiss, let me add, for him to learn political economy as well; he must know the military history of at least the last hundred years. But are military matters then exhausted by the army commander? No. Let us not forget the soldier, the individual platoon commander among whom military matters rest on the plane of a craft skill.

If a soldier doesn't know his maneuver, then he is simply cannon fodder; if he does know it, then he is a "craftsman." Beyond this what you have is already an art which utilizes the methods and conclusions of many sciences, employed in military matters. For example, methods of geography can and must be utilized for military affairs. A knowledge of statistics is absolutely required. Ethnography is required. So is history. All these are sciences. But the military business itself is not a science. We must distinguish, on the one hand, between science which establishes *the lawfulness of phenomena,* their causality and art, on the other—an art which has in view the *expediency of devices.* The expediency of devices, habits and methods and the lawfulness of objective phenomena—these are not one and the same thing. I am better able to elaborate an expedient method, the better I am familiar with the lawfulness of events; but it is nevertheless impermissible to confuse the latter with the former.

False Objections

Our military method in the Soviet Republic is determined in the last analysis by our technology, class correlations, and so on. But from these correct Marxist postulates it is impossible to deduce the subdivisions of a cavalry regiment. Gleb Uspensky depicted exquisitely in his story, "The Land's Power," how a peasant's entire life and all his thoughts are under the sway of the land and are wholly determined by the condition of the peasant's productive means. Marxism can supply an answer to the question: Why will the *moujik* continue to believe in hobgoblins so long as he goes around in *lapti*? *Lapti* (bast shoes) derive from and are determined by the peasant method of production; the latter also calls forth a whole number of other phenomena which are inseparable from the *lapti*: a narrow horizon, a slavish dependence on rain, sun and other elementary manifestations of nature; and all this, in the aggregate, creates the peasant's prejudices. Marxism can analyze and explain all this. But can Marxism teach how to make *lapti*? No, it can't. It can *explain* why the *moujik* goes around in *lapti*—because there is the forest, there is wood bark and poverty all around—but it is impossible to make *lapti* by means of the Marxist method! Comrade Akhov, however, wants to make *lapti* with the aid of Marxism. Nothing will come of it.

One speaker protested against calling military affairs an art on the ground that military affairs are not subject to the criterion of beauty. But this is already sheerest misconception. Trading is most surely not subject to esthetic criteria; there exists, nevertheless, the art of trading. Trade has its own complex methods, bound up with certain theories akin to science: Italian double bookkeeping, commercial correspondence, commercial geography, etc. What is trade—a science or an art? Marx made a science out of trade—in the sense that he established the laws of capitalist society, he made trade the object of scientific investigation. But can one trade "according to Marx"? No, this is impossible. One of the most stable, if not eternal principles of trade is the rule: "No cheating, no sale." Marxism explains whence arose this "principle" and how it later came to be supplanted by Italian double bookkeeping, which expresses the self-same thing but in a more delicate way. But is Marxism able to create a new system of bookkeeping? Or is a Marxist freed of the necessity of studying bookkeeping if he seriously wishes to take up trading? Behind the attempts to proclaim Marxism as the method of all sciences and arts there frequently lurks a stubborn refusal to enter new fields. For it is much easier to possess a *"passe-partout,"* that is, a master key that opens all doors and locks, rather than study bookkeeping, military affairs, etc. This is the greatest danger in all attempts to invest the Marxist method with such an absolute character. Marx attacked such pseudo-Marxists. In one of his letters he literally said, "I am no Marxist," when in place of an explanation of the historical process, in place of a careful and conscientious investigation of what was occurring Marx was proffered some kind of itinerary for history. Even less did Marx intend to replace all other fields of human knowledge by his social-historical theory. Does this mean that a military leader has no need of the Marxist method? Not at all. It would be absurd to deny the great importance of materialism for disciplining the mind in all fields. Marxism, like Darwinism, is the highest school of human thought. Methods of warfare cannot be deduced from Darwin's theory, from the law of natural selection; but an army leader who studied Darwin would be, given other qualifications, better equipped. He would have a wider horizon and be more fertile in devices; he would take note of those aspects of nature and man which previously had passed unnoticed. This applies to Marxism even to a greater extent.

The Province of Historical Analysis

One more comment on Comrade Akhov's remark concerning the role of historical analysis in clarifying this or that concept or hypothesis. It is absolutely correct that a historical point of view is fruitful in the extreme and that a history of science is superior to any Kantian epistemology. Man must keep cleaning his concepts and terms like a dentist cleans his instruments. But what we need for this is not a Kantian epistemology which takes concepts as being fixed once and forever. Terms must be approached *historically*. But a history of terms, hypotheses and theories does not replace science itself. Physics is physics. Military affairs are military affairs.

Marxism may be applied with the greatest success even to the history of chess. But it is not possible to learn how to play chess in a Marxist way. With the aid of Marxism we can establish that there once was an old *Oblomovist* nobility too lazy even to play chess; later, with the growth of cities, intellectuals and merchants appeared on the scene and there also arose the need of exercising the brains by playing checkers and chess. And now in our country workers go to chess clubs. Workers now play chess because they have overthrown those who used to ride on their backs. All this can be excellently explained by Marxism.

It is possible to show the entire course of the class struggle from one angle—that of the history of development of chess play. I repeat that it is possible by using the Marxist method to write an excellent book on the history of the development of chess play. But to learn to play chess "according to Marx" is impossible. Chess play has its own "laws," its own "principles." To be sure, I recently read that in the Napoleonic epoch chess play was maneuverist in character and so remained until the middle of the nineteenth century; whereas, during the interval of armed peace—from the Franco-Prussian war to the recent imperialist war—chess play remained wholly "positional," and nowadays it is allegedly again becoming fluid, "maneuverist." At all events this assertion is made by one American chess player. It is possible that social conditions, in some unknown ways, penetrate into the brains of a chess player and without being conscious of it, he reflects these conditions in his style of play. A materialist psychologist might find this of great interest. However, to play chess "according to Marx" is altogether impossible, just as it is impossible to wage war "according to Marx." Marxism does not teach how to use surprise when this becomes necessary in relation to the elusive Makhno.

What constitutes the essence of military matters is the to-

tality of rules for conquering. These rules are summed up
for better or for worse, in our statutes. Are they a science?
I think that our statutes cannot be called a science. They are
a system of prescripts, a body of rules and methods of a
craft or an art.

"Eternal Principles"

To those comrades who wish to build in military affairs
by means of the Marxist method I recommend that they review
our field statutes in this light and indicate just what changes
—from the standpoint of Marxism—should be introduced into
the rules for the gathering of intelligence, for securing one's
lines, for artillery preparation, or for attack. I should very
much like to hear at least a single new word in this sphere
arrived at through the Marxist method—not just "an opinion
or so" but something new and practical.

Such are the mistakes of young and immature Marxist
thought in the field of military theory. As against them there
are the mistakes of military academicians and metaphysicians
who tell us that military science discovers and formulates the
eternal principles in military matters. What do these principles signify? Are they scientific generalizations or practical
precepts? In what sense can they be called eternal?

War is a specific form of relations between men. In consequence, war methods and war usages depend upon the anatomical and psychical qualities of individuals, upon the form
of organization of the collective man, upon his technology, his
physical and cultural-historical environment, and so on. The
usages and methods of warfare are thus determined by changing circumstances and, therefore, they themselves can in nowise be eternal.

But it is quite self-evident that these usages and methods
contain elements of greater or lesser stability. Thus, for example, in cavalry methods are to be found elements in common
between ourselves and the epoch of Hannibal, and even earlier.
Methods used in aviation obviously are only of recent origin.
In our infantry methods are to be found traits in common with
the behavior of the most backward and primitive clans and
tribes who waged war against one another before the domestica-
tion of the horse. Finally, in military operation it is possible
in general to find the most elementary usages, common to men
and other fighting animals. Clearly, in these cases, too, it is
a question not of "eternal truths" in the sense of scientific

generalizations which derive from the properties of matter but of more or less stable usages of a craft or an art.

An aggregate of "military principles" does not constitute a military science, for there is no more a science of war than there is a science of locksmithing. An army leader requires the knowledge of a whole number of sciences in order to feel himself fully equipped for his *art*. But military science does not exist; there does exist a military craft which can be raised to the level of military art.

A scientific history of warfare is not military science but social science, or a branch of social science. A scientific history of warfare explains why in a given epoch, with a given social organization, men waged war in a certain way and not differently; and why such and such usages led in this epoch to victory whereas other methods brought defeat. Beginning with the general condition of productive forces, a scientific history of war must take into account all the superstructural factors, even the furthest removed, including the plans and the mistakes of the commanding staff. But it is quite self-evident that a scientific history of war aims by its very nature to explain that which undergoes change and the reasons for these changes, but not to establish eternal truths.

What truths can history give us? The role and significance of the growth of medieval cities in the development of military affairs. The invention of firearms. The overthrow of the feudal system and the significance of this revolution with respect to the army, and so on.

Marxist political economy is an incontestable science; but it is not a science of how to manage a business, or how to compete on the market, or how to build trusts. It is the science of how in a certain epoch certain economic relations (capitalist) took shape, and what conditions these relations internally, and constitutes their lawfulness. Economic laws established by Marx are not eternal truths but characteristic only of a specific epoch of mankind's economic development; and, in any case, they are not eternal principles as is represented by the bourgeois Manchester school, according to which private ownership of the means of production, buying and selling, competition and the rest are eternal principles of economy, deriving from human nature (about which however there is absolutely nothing eternal).

Source of the Blunder

Wherein lies the fundamental theoretical error of the liberal Manchester school of political economy? In this, that the generalizations (laws) which determine the economic practice of mankind in the epoch of commodity economy are transformed by the Manchester school into eternal principles which must serve eternally to guide economic activity.

Naturally, even for the Manchester economists it is no secret that the principles of commerce and competition did not always exist but arose at a certain stage of development. The doctrinaires of Manchesterism, however, get out of this difficulty by dating economic science from the origin of capitalist relations. Mankind has hitherto wallowed in the mire of dark ignorance or of feudal barbarism but later the truth of free trade was discovered, and this truth remains the eternal principle of human progress. For the Manchesterites, their economic laws possess the same significance as the laws of chemistry. In the Middle Ages mankind wallowed in the mire of serfdom, particularism and religious prejudices; neither the laws of chemistry nor the laws of the free market were known; later, both the former and the latter were discovered. Their objective value, their "eternity" is not compromised by the fact that people did not know about them earlier.

Doctrinaires in military affairs behave in exactly the same way with regard to military truths. Military generalizations, or more correctly the usages of a certain epoch, are transformed by them into eternal truths. If people were previously unaware of these eternal truths, so much the worse for those who wallowed in the mire of barbarism. But ever since their discovery, they remain eternal principles of military affairs. The erroneousness of such an approach becomes quite apparent if we use a proper scale for our inquiry. Medieval economy was not at all a product of ignorance; it had its own inner lawfulness derived from the then existing condition of human technology and the respective class structure of society.

The very simple laws which determined the economic relations between a feudal lord or seignior and his peasants, or a guild craftsman and his customers are just as "lawful" from the standpoint of economic science as the most complex laws of capitalist economy; both the former and the latter are transitional in character.

The army of *landsknechts*, the regular armies of the seventeenth and eighteenth centuries, the national army called to life by the Great French Revolution—all these correspond to definite epochs of economic and political development, and they all rest upon a certain technology on which they depend for their structure and methods of operation. Military history can and must establish this social conditioning of the army and its methods. But what does military philosophy do? As a rule it looks upon the methods and usages of a preceding epoch as eternal truths, at last discovered by mankind and destined to retain their meaning for all times and all peoples. The discovery of these eternal truths is linked primarily with the Napoleonic epoch. The same truths and principles are then discovered in the operations of Hannibal and Caesar. The period of the Middle Ages is turned into a hiatus in the course of which the eternal principles of war were forgotten along with the science and philosophy of antiquity.

Peculiarity of Military Affairs

There is, however, a difference between the mistakes of Manchesterism and the mistakes of the doctrinaires of eternal principles of military science. This difference lies in the difference between the two kinds of activity. Economic relations in capitalist society take shape, as Marx said, behind people's backs, arising from their ant-like economic labors; and the people then find themselves confronted with already crystallized property relations which determine the relations between man and man.

In military affairs the element of planned construction, of conscious direction by the human will comes into play on a far greater scope. Under capitalist relations plan, will, calculation, supervision, initiative are applied within the limits of an individual economy; and the laws of capitalist economy grow out of the relations between these individual economies: that is why they take shape "behind the backs" of people. But the army is by its very nature an all-state enterprise and consequently plans and projects are here applied within a state framework. This does not of course cancel the decisive dependence of military matters upon economy, but the subjective element in the person of military leaders attains a scope which cannot obtain in the sphere of economy.

This distinction, however, is by no means unconditional or unalterable in character. The action of the "eternal" principle of free competition led, as is well known, to monopoly, to the creation of powerful national and even international trusts. Individuals at the head of these trusts gain a field for strategical maneuvers wholly comparable to the theater of military activities during the last great war. Naturally, Rockefeller's arena for manifesting his "free will" in the domain of economic construction is far greater than was the case with some big industrialist or merchant 50 or 100 years ago. Rockefeller, however, is not an arbitrary violation of Manchesterite truths but their historical product, and at the same time their living negation.

Every industrialist-merchant, beginning with Gogol's Goat Beard and ending with the clean shaven Rockefeller, has his petty eternal truths of commercial operations: from "no cheating, no sale," and all the rest up to the complex calculations of an oil trust. Italian bookkeeping is of course not a science but an aggregate of commercial usages. It can be raised to the level of an art when applied along the proportions of a gigantic trust. The usages and habits of directing an industrial enterprise, the methods of supplying it with raw materials, the Taylor methods of labor organization, the methods of calculating prices, etc., represent a most complex practical system, which might even be called a "doctrine," in the sense of an aggregate of habits, usages, methods and means which best assure the plundering of the market. But of course this is not a science. To put it simply, political economy, that is, a genuine science, studies the internal relations of capitalist society but does not at all point out ways and means of surest enrichment. Military history, scientifically grounded, studies the typical traits of army and war organizations in each given epoch in correlation with the social structure of a given society, but does not and cannot at all teach how artillery is created and how conquest may be gained most surely.

Marshal Foch and Military Art

The military art of our time is summed up in statutes. These statutes are the concentrated experience of the past coined into currency intended for future use. This is an aggregate of the precepts of a craft or an art. Just as a collection of textbooks on the best organization of industrial enterprises, on calculation, on bookkeeping, on commercial correspondence and the rest does not comprise the science of capitalist society, so a collection of military manuals, regulations and statutes does not constitute military science.

In order to convince ourselves of the great unclarity and contradictoriness of the so-called eternal military principles (these are likewise the laws of military science) let us take the book "On the Principles of War," written by the outstanding victorious army leader in our time, Foch.

In his 1905 introduction Foch, on the basis of the initial data relating to the Russo-Japanese war, writes: "In the long run maneuverist offensive operations overcome any and all obstacles." Foch offers this idea as one of the eternal truths of military art, in contrast, let me add, to our native innovators who perceive in offensive maneuverist strategy qualities specific to revolutionary warfare. As we shall presently see, both sides are mistaken—both Foch who holds offensive maneuvering to be an eternal principle as well as those comrades who see in the maneuverist offensive the specific principle of the Red Army. In the introduction to the first edition of his book Foch approvingly cites the words of von der Golz:

> "While it is true that the principles of military art are eternal, the facts analyzed and taken into account by it are subject to constant evolution. Military theory is precisely comprised by a totality of these eternal principles."

The existence of this theory is just what makes, according to Foch, an art of war. One can thus say that military theory is constituted by a totality of those principles which were applied in all the correct operations, which when violated led to failure, and which must be applied in all the wars of the epochs to come. There thus exist such principles ("eternal" ones) as formed the foundations of military operations during the capture of Troy, when crafty Greeks hid in the belly of the wooden horse, just as they do in our time when a squadron of planes unloads hundreds of pounds of the most destructive explosives, or volumes of poison gases upon cities. What sort of principles are these?

Anatomical or psychological laws are not involved here. Unquestionably, there have been no very drastic changes in this connection. A Greek or Trojan whose heart was pierced dies just like one of our fighters. Cowards take fright and flee from battle. An army leader encourages his warriors, and so on. Man's basic psycho-physiological and anatomical structure has not altered very radically. Needless to say, the laws of nature have remained the same. But the relations between man and nature have altered in the extreme. The artificial milieu—weapons, instruments, machines—interposed by collective man between himself and nature has grown to such a degree as to completely transform his working habits, the organization of labor, the social relations. Since the days of Troy there has undoubtedly been preserved the urge in human groups (nations, classes) to destroy, conquer and subjugate one another. The artificial milieu, or human technology, in the broad meaning of the word, has transfigured war just as all other human relations. It is indubitable that even in the period of the siege of Troy this goal was already being attained not by means of nails and teeth alone but with the aid of artificial weapons interposed by man between himself and his enemy. This most common ground remains unchanged. In other words, war is a hostile clash between human groups equipped with the instruments for killing and destroying with **the direct aim of** gaining physical domination over the hostile side.

Foch's Principles

The concept of war is delimited in such a definition by social and historical frameworks. An outline of the general traits of war—first, the clash between human groups; second, the use of weapons; third, the goal of gaining preponderance over the hostile side—still does not, naturally, provide any principles of military art. At the same time, such a definition puts limits on the "eternity" of war itself. During that period when man had not yet learned to fight with clubs and stones, and not yet organized correctly acting herds (gens and tribes), there could obviously be no talk of war. For a clash between two of our distant progenitors biting through each other's throats for the sake of a female in the forest cannot be referred to as military art, bathed in the light of "eternal principles." The eternity of military art must thus at once be limited, and a running account opened for it only from the moment when man stood firmly erect on his hind legs, armed himself with a club and learned in battle, as in economic life, to act collectively, in detachments, although still without firmly established subdivisions.

Von der Goltz, and after him Foch, acknowledged that the factors studied by military art undergo change (the club, the musket, the automatic rifle, the machine gun, the cannon, and so on), but that the principles of the art remain if not eternal, then in any case unaltered since war first began.

What then are these principles? In his introduction to the second edition Foch seems to sponsor maneuverist offense as the main principle. But in the very first lecture he gives the following answer:

> "And so, the theory of war exists. It puts to the fore the following principles:
>
> "The principle of economy of forces.
> "The principle of freedom of action.
> "The principle of free disposition of forces.
> "The principle of security."

And so on.

And further, in order to bolster himself up ("comfort me in my disbelief"), Foch adduces a few citations, including the words of Marshal Bugeaud: "Absolute principles are few, but they nevertheless obtain."

Economy of Forces

But what comprises the first of these absolute principles, namely the principle of the economy of forces? The task of war is to overwhelm the enemy's living forces. This can be achieved only by means of a blow. For this blow a concentration of one's own forces is required. But before this blow can be dealt, it is necessary to discover the enemy's location, safeguard oneself against a sudden blow from his side, assure communications, and so on. This requires a disposition of corresponding detachments (reconnaissance, defense guards, etc.). The principle of economy of forces consists in assigning for auxiliary and preparatory tasks from among the basic detachments such forces—no more, no less—as are required by the very nature of these tasks; and at the same time, of assuring oneself at the decisive moment the possibility of bringing into play these auxiliary detachments in order to deal a concentrated blow. Foch explains that this result can be obtained only through the maneuverist offense of the basic army core as well as of the auxiliary detachments. The eternal principle of economy of forces is thus, according to Foch, characteristic only of maneuverist strategy. And it is hardly surprising that Foch permits into the holy of holies of military art only maneuverist offensive operations, holding that "theories previously current among us are false." Proceeding from maneuverist offense as the sole strategy, Foch predicts that the "initial combat actions will prove decisive in the next war." (Page 10.) In harmony with this same view, Foch draws the "conclusion that it [the next war] cannot be of long duration, and must be conducted with fierce energy and brought swiftly to its goal—otherwise it will be without results." (Page 38.)

In essence, it suffices to cite these conclusions in order for Foch's eternal principles to appear before us quite pathetically in the light of subsequent events. In the course of the last war the French army—after initial and costly attempts at offensives—went over to positional defense; the initial reverses did not at all predetermine the war's outcome as Foch had predicted; the war lasted four years; in essence, the war preserved throughout a positional character and was settled in the trenches; the first maneuverist period in the field served only to disclose the need of digging into the earth; the final period of field operations revealed only what had already been achieved in the trenches: the exhaustion of Germany's power of resistance.

This experience is of a certain value. If, according to Foch, the theories that dominated the French military school up to 1883 were false and the light of true principles began to dawn only toward the end of the last century, then a decade after his book was written it was already disclosed that the war had unfolded in complete contradiction to those predictions which Foch had deduced from eternal principles.

One might of course say that the error here is wholly on the side of Foch, who simply proved incapable of drawing the necessary conclusions from correct principles. But as a matter of fact, if the "eternal" principle of economy of forces is stripped of Foch's incorrect conclusions, then not much remains of the principle itself. According to Foch's line of thought, which is here nourished in the main by the Napoleonic experience, it is necessary first of all to locate the enemy, safeguard oneself by bringing up necessary reconnaissance and defense units to the front, along the flanks and in the rear; and then, having outlined the basic direction of the blow, to subordinate all forces to a single overwhelming offensive action. Essentially, the bare principle of "economy" of forces has nothing to do with all this. It all comes down to the pattern of the Napoleonic offensive maneuver in which all other considerations are subordinated to the moment of the concentrated blow.

The principle of economy of forces thus consists in an expedient distribution of forces between the basic and auxiliary units, all the while retaining the possibility of using all of them for the destruction of the enemy's living forces. However, the same Foch, basing himself on a famous conversation between Bonaparte and Moreau, gives another, more concrete and partial interpretation to the principle of economy of forces.

A Second Interpretation

On returning from Egypt Bonaparte explained to Moreau how he had secured himself a superiority of forces in the face of numerical inferiority by first descending with all his forces upon a single flank, smashing it and utilizing the ensuing confusion in order to strike with all his forces at the other flank. Does this mean that from the "theorem" (the expression is Foch's) of economy of forces is to be derived the principle of successive annihilation of the flanks? Obviously, no. We have here a specific case of a successful operation which is characterized by many most important elements: the number of troops, their armament, their respective mood, their disposition, the command, etc. In the concrete circumstances the problem was solved by Napoleon through one of several possible methods. Its successful outcome proves that Napoleon had the ability in the given instance of employing his forces; or, if you prefer, he used them economically; or he had applied the principle of "economy of forces." And nothing more.

But to interpret the principle of economy of forces in this way is only to give another name to the principle of expediency. This principle counsels us to act rationally, not to expend forces in vain. This smacks a little of—the "principles" of Kuzma Prutkov. If I remain ignorant of military affairs as such then this principle will afford me nothing. With a mathematical law which states that the square of the hypotenuse is equal to the sum of the squares of the other two sides, I can confront every corresponding phenomenon and apply the theorem practically. But if all I know is the "principle of economy of forces," what can I do with it? It is only a mnemonic sign which can be of use only after one possesses all the corresponding practical knowledge and habits. Surprise, economy of forces, freedom of action, initiative, and so on and so forth—these are only mnemonic signs for someone learned in military affairs. "Free masons" turned the signs of the mason's craft into freemasonic signs. Similarly, in military affairs a certain accumulated experience has a symbolic conditional denomination, that is all. There is nothing more.

Foch proves the absolute or eternal character of the principle of "freedom of action" by tracing it back to Xenophon:
"Military art consists in an ability to retain freedom of action." But what is the content of this freedom? First of all,
freedom of initiative must be maintained as against the enemy,
that is, he must not be given the opportunity to bind your will.
In this general form the principle is quite incontestable. But
it applies equally to fencing and to chess and generally to
all forms of sport which involve two sides, and finally, to
parliamentary and juridical debates. Foch later gives another
interpretation to this same principle. Freedom of action is
retained only by the commander-in-chief. All the other commanders are bound for they must act within the framework
of his assignments. Consequently, their will is placed under
the restraint not only of material circumstances, but also of
formal prescriptions. But economy of forces, or common sense,
or expediency—whichever you please—demands that the highest command not fix too narrow a framework for its subordinates. In other words, it is necessary to set a clearly defined
goal, leaving to the subordinate command the maximum freedom of choosing and combining means for the realization of
the set goal. In such a general form the principle is again
incontestable. The difficulty in issuing orders, however, lies
in finding that limit beyond which the definition of the desired
goal already passes into inordinate supervision over the choice
of means. The "theorem" does not in and of itself provide
any ready-made solution here. At best it serves only to re-
mind the commander that he must find some solution to this
problem.

But even apart from all this, it is quite clear that Foch
gives an equivocal interpretation to the principle of freedom
of action: On the one hand, it is that degree of initiative in
battle which assures the necessary independence from the enemy's will; and on the other hand, it is a sufficiently wide
freedom of maneuver for the lower command, within the limits of the goals and tasks fixed by the highest command.

War Is an Art

Neither the former nor the latter interpretation can, however, be called a theorem, even in the broadest meaning of
the word. In mathematics we understand by a theorem a
correlation of variable magnitudes that holds good under all

quantitative changes of these magnitudes. In other words, the equality is not disrupted by whichever arithmetical figures are substituted for the algebraic terms, designating the magnitudes. But what does the principle of economy of forces signify? Or the principle of freedom of action? Is this truly a theorem which permits, through a substitution of concrete magnitudes, of drawing correct practical conclusions? In no case. Any attempt actually to invest such a principle with "absolute" meaning, that is, raise it to the degree of a theorem, results in vacuities like: It is necessary to use all forces expediently; it is necessary to retain initiative of action; it is necessary to issue expedient or realizable orders, and therefore exclude from them superfluous conditions, and so on. In such a form these are not at all military principles, *but axioms of all purposive human activity*.

But, in point of fact, among military theoreticians these and similar principles are given a far more concrete interpretation. That is, these principles are (either openly or surreptitiously) made to include regiments, corps and armies of a specific structure and armament, which operate on a basis of numerous statutes and regulations, summing up the experience of the past. In such a form there is nothing eternal about these eternal principles; and they in nowise resemble theorems, but are conditional denominations of certain methods, empirical habits, positive and negative experiences, etc., etc.

In the nature of things, all military theoreticians cannot escape from the following contradiction: In order to demonstrate the eternal character of the principles of military art they have to throw out the entire "ballast" of living historical experience and reduce them to pleonasms, commonplaces, Euclidian postulates, logical axioms, etc. On the other hand, in order to demonstrate the importance of these principles in military affairs, they have to stuff these principles with the content of a specific epoch, a specific stage in the development of an army or in the development of military affairs; and thereby these principles are invested with the character of useful practical manuals for the memory.

These are not scientific generalizations but practical directives; not theorems, but statutes. They are not eternal, but transient. Their significance is all the greater, the less absolute they are, that is, the more they are filled with the concrete content of a given period of military affairs, its living peculiarities of organization, technique, and so on. They are not absolute but conditional. They constitute not a branch of science, but a practical manual of art.

An Abortive Attempt

Frederick the Great said: "War is a science for those who are outstanding; an art for mediocrities; a trade for ignoramuses." This statement is incorrect. There isn't and can't be a science of war, in the precise meaning of the word. There is the art of war. On the other hand, even a trade presupposes a schooling, and whoever has schooling is no ignoramus. It would be more correct to say that war is a skilled trade for the average individual and an art for an outstanding one. As regards an ignoramus, he is only the raw material of war; its cannon fodder, and not at all a skilled man.

The attempt to eternalize the Napoleonic principles proved, as we see, abortive. This was disclosed by the imperialist war. It could not have been otherwise, if only for the reason that the wars of the [French] revolution together with the Napoleonic wars that grew out of the former were distinguished by the colossal moral and political preponderance of the revolutionary French people and their army over the rest of Europe. The French took the offensive in the name of a new idea, closely bound up with the powerful interests of the popular masses. The opposing armies put up a half-hearted defense for the old system. But during the last imperialist war neither side was the bearer of a new principle embodied in a new revolutionary class. On both sides the war was imperialist in character, but, at the same time, the very existence of both sides, above all Germany and France, was equally threatened. There was no violent blow which would have immediately caused demoralization and dejection in the opposing camp; nor could such a blow have been struck in view of the great human and material strength of both camps who moved up all their forces and resources gradually.

For this reason the initial battles, in contrast to Foch's forecasts, did not at all predetermine the outcome of the war. For this very reason, offensives were shattered by counteroffensives and the armies, leaning more and more on their rear, dug into the earth. For this very same reason, the war lasted a long time—until the moral and material resources of one side were exhausted. The imperialist war thus went its course from beginning to end in violation of the "eternal" maneuverist offensive principle proclaimed by Foch. This circumstance is underscored all the more by the fact that Foch turned out to be the victor, despite and against his own principles. The explanation for this is to be found in the fact that while Foch's principles were against him the English and American soldiers, and especially Anglo-American munitions, tanks and planes, were with him.

One may of course say that the principle of economy of forces remains valid for positional warfare as well. For in this case, too, an expedient distribution of forces between frontal detachments and the various categories of reserve is required. This is quite indisputable. But in such a general presentation, not even a trace remains of the scheme whereby forces are distributed for a concentrated offensive blow. The "eternal" principle dissolves into a commonplace. In positional, defensive, offensive, as well as maneuverist wars it is necessary to have an expedient and economic distribution of forces depending upon the task at hand. It is quite self-evident that this "eternal principle" applies to industry and commerce as much as to war. It is always necessary to utilize one's forces economically, that is, obtain maximum results from a minimum expenditure of energy. All human progress, and first of all, technology are based on this "eternal" principle. Man began to use a stone ax, a club, etc., because he thus obtained the greatest results with the least expenditure of effort. Precisely for this same reason man went from the club to the spear and the sword. From them—to the gun and the bayonet, and later to the cannon, etc. For the very same reason, he now passes to the electric plow. The eternal principle of war thus comes down to a "principle" which is the motor of all human development. As regards the concrete interpretation given by Foch to the principle of economy of forces, it proved to be an abortive attempt to give an absolute character to the Napoleonic offensive maneuver which is resolved by a concentrated blow.

A Materialist Approach

And so, insofar as the principle of economy of forces is "eternal," it contains nothing military. And insofar as it is given a military interpretation, there is nothing eternal about it.

But why does all this talk about "eternal" principles continue to persist? Because, as has already been pointed out, at the basis there is man. Human qualities undergo little change. Anatomical, physiological, psychological qualities alter slowly as compared to changes of social forms. The relation of man's hands and feet and the structure of his skull in our epoch are approximately the same as in the days of Aristotle. We know that Marx used to read Aristotle with delight. And were it possible to assume Aristotle's transfer to our epoch in order for him to read Marx's books, then in all likelihood Aristotle would have understood them excellently.

Man's anatomical and psycho-physical make-up is far more stable than social forms are. Corresponding to this there are two sides in military affairs: There is the individual side, which finds its expression in certain habits and methods, determined to a large measure by the biological nature of man, not eternal but stable; and there is the collective-historical side which depends on the social organization of man in war. But it is precisely this latter moment which decides the issue, because war begins when socially organized armed man enters into combat with another socially organized armed man. Otherwise we would have a fight—between animals.

Comrade Lukirsky approached the question from the following standpoint: There is, on the one hand, experience and empirical inquiry—an imperfect method; and there is on the other hand, "pure reason" which deductively, by means of logical methods arrives at "absolute" deductions and thereby enriches military affairs. As a materialist I have become accustomed to look upon reason as one of the organs of historical man, developed in the process of man's adaptation to nature. I cannot oppose reason to matter; I cannot agree to think that reason can supposedly give birth to that which material experience has not already provided. Our reason only coordinates and correlates conclusions from our practice; from "pure" reason man can deduce nothing new, nothing he had not abstracted from experience. Naturally, experience does not "take shape" mechanically, but rather there is an order introduced into it—an order which corresponds to the order of the manifestations in themselves and which leads to the knowledge of the lawfulness of these manifestations. But to think that reason can arbitrarily give birth to a conclusion which is not prepared by and grounded in experience—this is absolutely wrong. And if that is the case, neither can there be principles of a twofold character: practical and eternal.

In conclusion, let me say that we have already had one discussion on the subject of "military doctrine," and now we have reached the ultimate philosophic heights. The time has come for us to begin the downward climb and get down to the tasks of practical schooling. We had once planned to put out *A Syllabus For An Individual Commander*, but nothing has yet come of this project. Which is more difficult to write—abstract theses or a syllabus for an individual commander? The latter is a hundred times more difficult; but, by way of compensation, it is a thousand times more fruitful.

I wish to utilize this large gathering, the presence of many competent workers, in order to make once again my proposal that we supply individual commanders with general directives —with a model little book "How To Conquer Knowingly." It would be an excellent school for all of us were we to set down our military experience in such clear and precise regulations that an individual commander could not only read but study them with profit.

Out of the very same bricks it is possible to build a factory, a home, or a temple. The only requirement is that the bricks be made of good material and properly baked. The very same regiments, with one and the same schooling, under one and the same objective circumstances can be deployed and utilized for the most diverse strategical and tactical assignments. The sole requirement is that the basic cell—*the subdivision*—be viable and resilient. And for this we need a conscious individual commander who knows his business and his own worth. Our task of tasks consists in educating such individual commanders. To educate the individual proletarian commander does not at all mean to implant in his mind the idea that hitherto there have allegedly been bourgeois tactics and now the time has come for proletarian tactics. No. Such an education would lead him astray. To create the individual proletarian commander means to assist our present individual commander in acquiring at the very least that sum of knowledge and habits which such an individual possesses in bourgeois armies in order that he may consciously use this knowledge and these habits in the interests of the working class.

5. MARXISM AND

MILITARY WARFARE

March 19, 1924

Friedrich Engels' book is, for the most part, an analytical chronicle of the Franco-Prussian War of 1870-71. It is composed of articles published in the English *Pall Mall Gazette* during the war events. This is enough to make it clear that the reader cannot count on finding in these articles a sort of monograph on war or any systematic presentation of the theory of the art of war. No, Engels' task consisted—proceeding from the general appraisal of the forces and means of the two adversaries and following from day to day the manner of employing these forces and means—in helping the reader orient himself in the course of the military operations and even in lifting the so-called veil of the future a little from time to time. Military articles of this kind fill at least two-thirds of the book. The remaining third consists of articles devoted to various special fields of the military profession again in closest connection with the course of the Franco-Prussian War: "How to Fight the Prussians," "The Rationale of the Prussian Army System," "Saragossa-Paris," "The Emperor's Apologia," and so on.

It is clear that a book of this kind cannot be read and studied like the other, purely theoretical, works of Engels. To understand perfectly the ideas and evaluations of a concrete, factual kind contained in this book, all the operations of the

---

*Friedrich Engels: **Notes on the War.** Sixty articles reprinted from the **Pall Mall Gazette**, 1870-71. Edited by Friedrich Adler. Vienna. 1923.

Franco-Prussian War must be followed step by step on the map, and the viewpoints set forth in the latest war-historical literature taken into consideration. The average reader cannot of course set himself the task of such a critical-scientific labor: it calls for military training, a great expenditure of time and special interest in the subject. But would such interest be justified? In our opinion, Yes. It is justified primarily from the standpoint of a correct evaluation of the military level and the military perspicacity of Friedrich Engels himself. A thorough examination of Engels' extremely concise text, the comparison of his judgments and prognoses with the judgments and prognoses made at the same time by military writers of the time, could count on attracting great interest, and would not only be a valuable contribution to the biography of Engels—and his biography is an important chapter in the history of socialism—but also as an extremely apt illustration in the question of the reciprocal relations between Marxism and the military profession.

A Thoroughgoing Work

Of Marxism or dialectics, Engels says not a word in all these articles; which is not to be astonished at, for he was writing anonymously for an arch-bourgeois periodical and that at a time when the name of Marx was still little known. But not only these outward reasons prompted Engels to refrain from all general-theoretical considerations. We may be convinced that even if Engels had had the opportunity then to discuss the events of the war in a revolutionary-Marxian paper—with far greater freedom for expressing his political sympathies and antipathies—he would nevertheless hardly have approached the analysis and the estimation of the course of the war differently than he did in the *Pall Mall Gazette*. Engels injected no abstract doctrine into the domain of the science of war from without and did not set up any tactical recipes, newly-discovered by himself, as universal criteria.

Regardless of the conciseness of the presentation, we see nonetheless with what attentiveness the author deals with all the elements of the profession of war, from the territorial areas and the population figures of the countries involved down to the biographical researches into the past of General Trochu for the purpose of being better acquainted with his methods and habits. Behind these articles is sensed a vast preceding and continuing labor. Engels, who was not only a profound thinker, but also an excellent writer, dishes up no raw material for the reader. This may give the impression of cursoriness in some of his observations and generalizations. This is not really so. The critical elaboration he made of the empirical material is tremendously far-reaching. This may be perceived from the fact that the subsequent course of the events of the war repeatedly confirmed Engels' prognoses. We need not doubt that a searching study of this work of Engels in the sense referred to by our young war theoreticians would show even more the great earnestness with which Engels treated the conduct of war as such.

Quantity and Quality in War

But even among those who merely read and do not study the book—and they will make up the overwhelming majority even among the military people—this work of Engels will arouse great interest, not because of its analytical presentation of the various military operations but because of the general appraisal of the course of the war and the judgments made in the specific military fields that are scattered through many passages of his war chronicle and in part, as already stated, are dealt with in entire articles.

The old idea of the Pythagoreans, that the world is ruled by numbers—in the realistic and not the mystical sense of the word—may be especially well applied to war. First of all—the number of battalions. Then the number of guns, the number of ordnance pieces are expressed quantitatively: through the range of the firearm, through its accuracy. The moral qualities of the soldiers are expressed in the capacity to endure long marches, to hold out for a long time under enemy fire, etc. However, the further we penetrate into this field, the more complicated the question becomes. The amount and character of the equipment depends upon the condition of the forces of production of the country. The composition of the army and the personnel of its command is conditioned by the social structure of society. The administrative supply apparatus depends upon the general-state apparatus, which is determined by the nature of the ruling class. The morale of the army depends upon the mutual relations of the classes, upon the ability of the ruling class to make the tasks of the war the subjective aims of the army. The degree of the ability and talent of the commanding personnel depends in turn upon the historical rôle of the ruling class, upon its ability to concentrate the best creative forces of the land upon their aims, and this ability depends again in turn upon whether the ruling class plays a progressive historical rôle or has outlived itself and is only fighting for its existence.

Here we have disclosed only the basic coördinates, and even these only schematically. In reality, the dependence of the various fields of war conduct upon each other, and of all of them taken together upon the various aspects of the social order, are much more complex and detailed. On the battlefield, this is all summed up in the last analysis in the number of ordinary soldiers, the commander, the dead and wounded, prisoners and deserters, in the size of the conquered territory and in the number of trophies. But how is the end-result to be foreseen? If it were possible exactly to register and determine in advance all the elements of a battle and a war, there would be no war altogether, for nobody would ever think of heading toward a defeat assured in advance. But we cannot talk of such an exact foreseeing of all the factors. Only the most immediate material elements of war may be expressed in numbers.

In so far, however, as it is a question of the dependence of the material elements of the army upon the economy of the country as a whole, any appraisal, and therefore also any foresight, will have a much more conditional value. This applies especially to the so-called moral factors: the political equilibrium in the country, the tenacity of the army, the attitude of the hinterland, the coördination of the work of the state apparatus, the talents of the commander, etc. Laplace says that an intellect that was in a position to take in at a glance all the processes developing in the universe would be able to foretell without error everything that would take place in the future. This undoubtedly follows from the principle of determinism: no phenomenon without a cause. But, as is known, there is no such intellect, neither individual nor collective. Therefore it is also possible for even the best informed and most gifted men to err very often in their foresight. But it is clear that the right foresight is most closely approached the better the elements of the process are known, the greater the ability to find their right place, to estimate them and combine them, the greater the scientific creative experience, the broader the horizon.

Infantry, Then and Now

In his military newspaper chronicle, so modest in the task it sets itself, Engels always remains himself: he brings to his

work the sharp eye of a military analyst and synthesizer who
has gone through the great social-theoretical school of Marx-
Engels, the practical school of the Revolution of 1848, and the
First International.

"Let us now compare the forces," says Engels, "that are
being got ready for mutual destruction; and to simplify matters, we will take the infantry only. The infantry is the arm
which decides battles; any trifling balance of strength in cavalry and artillery, including mitrailleurs and other miracle-
working engines, will not count for much on either side."
(*Notes on the War*, Note I, page 1.)

What was right, by and large, for France and Germany in
1870, would undoubtedly no longer hold for our time. It is
now impossible to determine the relationship of military forces
only by the number of battalions. It is true that the infantry
remains even today the main factor in battle. But the rôle of
the technical coefficients in the infantry has grown extraordi-
narily, although in very unequal measure in the different
armies: we have in mind not only the machine guns which
were still "miracle-working" in 1870; not only the artillery,
which has increased in number and importance; but also perfectly new auxiliaries: the motor truck for war as well as for
transportation purposes, aviation, and war chemistry. Any
statistics that do not take these "coefficients" into consideration and deal only with the number of battalions, would now
be completely unreal.

On the basis of his calculations, Engels reaches the conclusion: Germany has a far greater number of trained soldiers
at its disposal than France, and the superiority of the Germans
will manifest itself increasingly with time—unless Louis Napoleon forestalls the enemy at the very outset and strikes him
decisive blows before he can bring his potential superiority
into play.

Therewith Engels gets to strategy, to that independent domain of the highest war art which is, however, connected by
means of a complicated system of levers and transmissions
with politics, economics, culture and administration. With
regard to strategy, Engels deems it necessary to make the in-
escapable realistic restrictions right at the outset:

"In the meantime it is well to remember that these strategic plans can never be relied upon for the full effect of what
is expected from them. There always occurs a hitch here and
a hitch there; corps do not arrive at the exact moment when
they are wanted; the enemy makes unexpected moves, or has

taken unexpected precautions; and finally, hard, stubborn fighting, or the good sense of a general, often extricates the defeated army from the worst consequences a defeat can have —the loss of communication with its base." (*Ibid.*, Note III, page 6.)

This is indubitably correct. Only the late Pfuel or one of his belated admirers could raise objection to such a realistic conception of strategy: to take into account what is most important in the whole war plan and to do it with the greatest completeness permitted by circumstances; consideration for those elements which cannot be determined in advance; formulation of orders in such a flexible way as to make them adaptable to the actual situation and its unforeseen variants; and the main thing: timely recording of every essential change in the situation and corresponding alterations of the plan or even its complete rearrangement—this is precisely what the true art of the conduct of war consists of. If the strategical plan could be invested with an exhaustive character, if the state of the weather, of the soldier's stomach and legs, and the intentions of the adversary could be accounted for in advance, then any robot who has mastered the four first rules of arithmetic could be a victorious field commander. Luckily or unluckily, it cannot be done. The war plan has in no wise an absolute character, and the existence of the best plan, as Engels rightly points out, far from guarantees the victory. On the other hand, any lack of plan makes defeat inevitable. Any commander who is half-way serious knows the orienting, if not absolute, value of a plan. But the commander who would reject a plan for this reason, would either be shot or locked in a madhouse.

Politics in the Army

How did matters stand with the strategical plan of Napoleon III? We already know that Germany's vast potential superiority lay in the numerical preponderance of trained human material. As Engels emphasizes, Bonaparte's task consisted in making the employment of this superiority impossible by means of rapid, resolute attacks upon the enemy. One would think that the Napoleonic tradition would have favored precisely such a procedure. But the realization of such audacious war plans, disregarding everything else, depends also upon the exact work of the commissariat, and the whole régime of the Second Empire, with its unbridled and incompetent bureaucracy, was in no wise fit to assure the provisioning and equipping of the troops. Hence the friction and loss of time right at the beginning of the war, the general help-lessness, the impossibility of carrying out any plan, and as a result of all this—the collapse.

In some passages, Engels mentions fleetingly the harmful effect that the penetration of "politics" can have in the course of war operations. This observation of his seems at first blush to be in conflict with the conception that war, by and large, is nothing but a continuation of politics. In reality, there is no contradiction here. The war continues politics, but with special means and methods. When politics is compelled, for the solution of its fundamental tasks, to resort to the aid of war, this politics must not hamper the course of the war oper-ations for the sake of its subordinated tasks. When Bonaparte took actions which were obviously inexpedient from the mili-tary standpoint in order, as Engels opines, to influence "pub-lic opinion" favorably with ephemeral successes, this was un-doubtedly to be regarded as an inadmissable invasion of poli-tics into the conduct of the war which made it impossible for the latter to accomplish the fundamental tasks set by politics. To the degree that Bonaparte was forced, in the struggle to preserve his régime, to permit such an invasion of politics, an obvious self-condemnation of the régime was revealed which made the early collapse inevitable.

When the vanquished land, following the complete de-feat and capture of its armed forces, attempted under Gam-betta's leadership to establish a new army, Engels followed these labors with astonishing understanding of the essence of military organization. He characterized splendidly the young, undisciplined troops who had been assembled by im-provization. Such troops, he says, "are but too ready to cry '*trahison*' unless they are at once led against the enemy, and to run away when they are made seriously to feel that ene-my's presence." (*Ibid.*, pages 88*f*.) It is impossible not to think here of our own first troop detachments and regiments in 1917-18!

Popular Armed Forces

Engels has an excellent knowledge of where, given all the other necessary pre-conditions, the main difficulties lie in transforming a human mass into a company or a battalion. "Whoever," says he, "has seen popular levies on the drill-ground or under fire—be they Baden Freischaaren, Bull-Run Yankees, French Mobiles, or British Volunteers—will have perceived at once that the chief cause of the helplessness and unsteadiness of these troops lies in the fact of the officers not knowing their duty." (*Ibid.,* page 79.)

It is most instructive to see how attentively Engels treats the home guards of an army. How far removed this great revolutionist is from all the pseudo-revolutionary chatter which was very popular in France right at that time—on the saving power of a mass mobilization (*levée en masse*), an armed nation (armed in a trice), etc. Engels knows very well the great importance officers and non-commissioned officers have in a battalion. He makes exact calculations on what resources in officers have remained to the republic following the defeat of the regular forces of the Empire. He gives the greatest attention to the development of those features in the new, so-called Loire army which distinguish it from armed human mass. Thus, for example, he records with satisfaction that the new army not only intends to proceed unitedly and to obey orders, but also that it "has learned again one very important thing which Louis Napoleon's army had quite forgotten—light infantry duty, the art of protecting flanks and rear from surprise, of feeling for the enemy, surprising his detachments, procuring information and prisoners." (*Ibid.,* page 96.)

This is how Engels is everywhere in his "newspaper" articles: bold in his grasp of affairs, realistic in method, perspicacious in big things and little, and always scrupulous in the manipulation of materials. He counts the number of drawn and smooth-bore gun barrels of the French, repeatedly checks on the German artillery, thinks of the qualities of the Prussian cavalry horse, and never forgets the qualities of the Prussian non-commissioned officer. Faced in the course of events by the problem of the siege and defense of Paris, he investigates the quality of its fortifications, the strength of the artillery of the Germans and the French, and takes up very critically the question of whether there are regular troops behind the walls of Paris that may be called effective for battle. What a pity we did not have this work of Engels in 1918! It would surely have helped us overcome more speedily and easily the then widely disseminated prejudice with which it was sought to counterpose "revolutionary enthusiasm" and the "proletarian spirit" to a professional organization, flawless discipline and trained command.

Engels' Method

The military-critical method of Engels is very clearly expressed, for example, in his thirteenth letter, which deals with the rumor launched from Berlin about "a decisive ad-

vance upon Paris." The article on the fortified camp of Paris
(Letter Sixteen) met with Marx's enthusiastic applause. A
good example of Engel's treatment of military problems is
offered by the twenty-fourth letter, which deals with the siege
of Paris. Engels sets forth two fundamental factors in advance: "The first is that Paris cannot hope to be relieved, in
useful time, by any French army from without.... The second point settled is that the garrison of Paris is unfit to act on
the offensive on a large scale." (*Ibid.*, page 71.) All the other
elements of his analysis rest upon these two points. Very interesting are two judgments on the *franctireur* war and the
possibilities of employing it, a question which will not lose
its importance for us even in the future. Engels' tone gains
in confidence with every letter. This confidence is justified
inasmuch as it has been confirmed by a twofold test: on one
side, by comparison with what the "genuine" military people
have written on the same questions, and on the other, by a
more effective test—the events themselves.

Relentlessly ruling out of his analysis every abstraction,
regarding war as a material chain of operations, considering
every operation from the the standpoint of the actually existing forces, means and the possibility of employing them, this
great revolutionist acts as . . . a war specialist, that is, as a person who by mere virtue of his profession or his vocation proceeds from the internal factors of the conduct of war. It is
not astonishing that Engels' articles were attributed to re-
nowned military men of the time, which led to Engels' being
nicknamed the "General" among his circle of friends. Yes,
he handled military questions like a "general," perhaps not
without substantial defects in specific military domains and
without the necessary practical experience, but, in exchange,
with a talented head such as not every general has on his
shoulders.

But, it might be asked, where, after all this, is Marxism?
To this may be replied that it is precisely here—up to a certain degree—that it is expressed. One of the fundamental
philosophical premises of Marxism says that the truth is always concrete. This means that the profession of war and its
problems cannot be dissolved into social and political categories. War is war, and the Marxist who wants to judge it
must bear in mind that the truth of war is also concrete. And
this is what Engels' book teaches primarily. But not this
alone.

If military problems may not be dissolved into general

political problems, it is likewise impermissible to separate
the latter from the former. As we have already mentioned,
war is a continuation of politics by special means. This profoundly dialectical thought was formulated by Clausewitz.
War is a continuation of politics: whoever wishes to understand the "continuation" must get clear on what preceded it.
But continuation — "by other means" — signifies: it is not
enough to be well oriented politically in order to be able
therewith also to estimate correctly the "other means" of war.
The greatest and incomparable merit of Engels consisted in
the fact that while he had a profound grasp of the independent character of war—with its own inner technique, structure, its methods, traditions and prejudices—he was at the
same time a great expert in politics, to which war is in the
last analysis subordinated.

It need not be said that this tremendous superiority could
not guarantee Engels against mistakes in his concrete military judgments and prognoses. During the Civil War in the
United States, Engels overrated the purely military superiority that the Southerners displayed in the first period and was
therefore inclined to believe in their victory. During the
German-Austrian War in 1866, shortly before the decisive
battle at Königgrätz, which laid the foundation stone for the
predominance of Prussia, Engels counted on a mutiny in the
Prussian Landwehr. In the chronicle of the Franco-Prussian
War, too, a number of mistakes in isolated matters can undoubtedly be found, even though the general prognosis of
Engels in this case was incomparably more correct than in
the two examples adduced. Only very naïve persons can
think that the greatness of a Marx, Engels or Lenin consists
in the automatic infallibility of all their judgments. No, they
too made mistakes. But in judging the greatest and most complicated questions they used to make fewer mistakes than all
the others. And therein is shown the greatness of their thinking. And also in the fact that their mistakes, when the reasons
for them are seriously examined, often proved to be deeper
and more instructive than the correct judgment of those who,
accidentally or not, were right as against them in this or that
case.

Class Tactics and Strategy

Abstractions of all kinds, such as that every class *must*
have specific tactics and strategy peculiar to itself, naturally
find no support in Engels. He knows all too well that the

foundation of all foundations of a military organization and a war is determined by the level of the development of the productive forces and not by the naked class will. To be sure, it may be said that the feudal epoch had its own tactics and even a number of coördinated tactics, that the bourgeois epoch, in turn, has known not one but several tactics, and that socialism will surely lead to the elaboration of new war tactics if it is forced into the position of having to coëxist with capitalism for a long time. Stated in this general form it is correct, in the degree that the level of the productive forces of capitalist society is higher than that of feudal, and in the socialist society it will with time be still higher. But nothing further than this. For it in no wise follows that the proletariat which has attained power and disposes of only a very low level of production, can immediately form new tactics which—in principle—can only flow from the enhanced development of the productive forces of the future socialist society.

In the past we have very often compared economic processes and phenomena with military. Now it will perhaps not be without value to counterpose some military questions to the economic, for in the latter domain we have already garnered a fairly considerable experience. The most important part of industry is working with us under conditions of socialist economy, by virtue of the fact that it is the property of the workers' state and produces on its account and under its direction. By virtue of this circumstance, the social-juridical structure of our industry is incisively distinguished from the capitalistic. This finds its expression in the system of administration of industry, in the election of the directing personnel, in the relationship between the factory management and the workers, etc. But how do matters stand with the process of production itself? Have we perhaps created our own socialist methods of production, which are counterposed to the capitalistic? We are still a long distance from that. The methods of production depend upon the material technique and the cultural and productive level of the workers. Given the worn-out installations and inadequate utilization of our plant, the production process now stands on an incomparably lower level than before the war. In this field we have not only created nothing new, but we can only hope after a number of years to acquire those methods and means of production which are at present introduced into the advanced capitalist countries and which assure them thereby of a far higher productivity of labor. If, however, this is how matters stand in the field of economy, how can it be otherwise in principle in the military field? Tactics depend upon the existing war technique and the military and cultural level of the soldiers.

To be sure, the political and social-juridical structure of our army is basically different from the bourgeois armies. This is expressed in the selection of the commanding personnel, in the relationship between it and the soldier-mass, and primarily in the political aims that inspire our army. But in no wise does it follow from this that now, on the basis of our low technical and cultural level, we are already able to create tactics, new in principle and more perfected, than those which the most civilized beasts of prey of the West have attained. The first steps of the proletariat which has conquered power—and these first steps are measured in years— must not—as the same Engels taught—be confused with the socialist society, which stands on a higher stage of development. In accordance with the growth of the productive forces on the basis of socialist property, our production process itself will also necessarily assume a different character than under capitalism. In order to change the character of production qualitatively, we need no more revolutions, no shakeups in property, etc.: we need only a development of the productive forces on the foundation already created. The same applies also to the army. In the Soviet state, on the basis of a working community between workers and peasants, under the direction of the advanced workers, we shall undoubtedly create new tactics. But when? When our productive forces outstrip the capitalistic, or at least approximate them.

It is understood that in case of military conflicts with capitalist states, we have an advantage, a very small one but an advantage nonetheless, that may cost our possible enemies their heads. This advantage consists in the fact that we have no antagonism between the ruling class and the one from which the mass of the soldiers is composed. We are a workers' and peasants' state and at the same time a workers' and peasants' army. But that is no military superiority but a political one. It would be extremely unwarranted to draw conclusions from this political advantage that would lead to military arrogance and self-overestimation. On the contrary, the better we recognize our backwardness, the more we refrain from braggadocio, the faster we learn the technique and tactics of the advanced capitalist armies, the more warranted will be our hope that in the event of a military conflict we shall drive a sharp wedge, not only of a military but also of a revolutionary kind, right between the bourgeoisie and the soldier-mass of its armies.

I am not certain whether it is appropriate here to mention the famous discovery of the no less famous Chernov* on the "nationalism" of Marx and Engels. The book before us gives a clear answer to this question too, which does not alter our former judgment, but, on the contrary, strengthens it in the most striking way. The interests of the revolution were, for Engels, the highest criterion. He defended the national interests of Germany against the Empire of Bonaparte, because the interests of the unification of the German nation under the concrete historical relations of the time signified a progressive, potentially-revolutionary force. We are guided by the same method when we now support the national interests of the colonial peoples against imperialism. This position of Engels found its expression, and a very restrained one, in the articles of the first period of the war. How could it have been otherwise: It was after all impossible for Engels, just to please Napoleon and Chernov, to evaluate the Franco-Prussian War in opposition to its historical meaning only because he was himself a German. But the minute the progressive historical task of the war was achieved, the national unification of Germany assured, and besides this, the Second Empire overturned —Engels radically changed his "sympathies"—if we may express his political tendency by this sentimental term. Why did he do this? Because it was now a question, beyond what was achieved, of assuring the predominance of the Prussian Junker in Germany and of Prussianized Germany in Europe. Under these conditions, the defense of dismembered France became a revolutionary factor or it might have become one. Engels stands here entirely on the side of the French struggle of defense. But just as in the first half of the war, he does not permit his "sympathies"—or at least endeavors not to permit them—to gain influence over the objective evaluation of the war situation. In both periods of the war, he proceeds from a consideration of the material and moral war factors and seeks a firm objective basis for his prognosis.

It will not be superfluous to point out, at least cursorily, how the "patriot" and "nationalist" Engels, in his article on the fortification and defense of the French capital, sympathetically considered the possibility of an English, Italian, Austrian and Scandinavian intervention in favor of France. His arguments in the columns of an English paper are noth-

*Chernov was the outstanding leader of the Social-Revolutionary Party of Russia, a petty bourgeois, non-Marxian organization.—Trans.

ing but an attempt to promote the intervention of a foreign power in the war against the dear Hohenzollern fatherland. This certainly weighs much heavier than even a sealed railway car!*

Engels' interest in military questions had not a national but a purely revolutionary source. Emerging from the events of 1848 as a mature revolutionist who had the *Communist Manifesto* and revolutionary struggles behind him, Engels regarded the question of the conquest of power by the proletariat as a purely practical question, whose solution depended not least of all upon war problems. In the national movements and war events of 1859, 1864, 1866, 1870-71, Engels sought for the direct levers for a revolutionary action. He investigates every new war, discloses its possible connection with revolution, and seeks for ways of assuring the future revolution by the power of arms. Herein lies the explanation for the lively and active, by no means academic and not merely agitational treatment of army and war problems that we find in Engels. With Marx, the position in principle was the same. But Marx did not occupy himself specifically with military questions, relying entirely on his "second fiddle" in such matters.

In the epoch of the Second International, this revolutionary interest in war questions, as, moreover, in many other questions, was almost completely lost. But opportunism was perhaps most plainly expressed in the superficial and disdainful attitude toward militarism as a barbaric institution unworthy of enlightened social-democratic attention. The imperialist war of 1914-18 recalled to mind again—and with what implacable inconsiderateness!—that militarism is not at all merely an object for stereotyped agitation and speeches in Parliament. The war took the socialist parties by surprise and converted their formally oppositional attitude toward militarism into humble genuflections. It was the October Revolution that was first called upon not only to restore the active-revolutionary attitude toward war questions, but also to turn the spearhead of militarism practically against the ruling classes. The world revolution will carry this work to the end.

---

*An allusion to the sealed railway car in which Lenin, together with other Bolshevik and Menshevik leaders, travelled through Germany, by arrangement with the Hohenzollern government in 1917, in order to reach revolutionary Russia. The "sealed car" episode was used by Russian reactionaries, and even some "socialists," as the basis for a slander campaign against Lenin as a "German agent."—Trans.

Appendix One

SCIENTIFICALLY —OR "SOMEHOW"?

A Letter to a Friend

January 10, 1919

DEAR FRIEND,

You ask how it can have happened that the question of specialists, such as the officers of the old General Staff, has assumed such great importance among us. Let me tell you that what is at issue here is actually not the matter of military specialists—it is a question both broader and deeper than that.

We are the party of the working class. Together with its advanced elements we spent decades in underground conditions, carried on our struggle, fought on the barricades, overturned the old régime, cast aside all the in-between groups such as the Socialist Revolutionaries and Mensheviks and, at the head of the working class, took power into our hands. But though our party is deeply and unbreakably linked with the working class it has never been and cannot become a mere flatterer of the working class, expressing a gratification with whatever the workers may be doing. We treated with contempt those who preached to us that the proletariat had taken power 'too soon', as though a revolutionary class can take power whenever it likes and not when history forces it to take power. But at the same time we never said, and we do not say now, that our working class has attained full maturity and can cope, as though playing a game, with all tasks and resolve all difficulties. The proletariat and, all the more so, the peasant masses, have only recently emerged, after all, from many centuries of slavery and bear all the consequences of oppression, ignorance and darkness. The conquest of power, in itself, does not at all transform the working class and does not confer upon it all the attainments and qualities it needs: the conquest of power merely opens up for it the possibility of really studying and developing and ridding itself of its historical shortcomings.

By a tremendous effort the upper stratum of the Russian working class has accomplished a gigantic historical task. Even in this upper stratum, however, there is still too much half-knowledge and half-skill, too few workers who, by virtue of their knowledge, breadth of horizon and energy are capable of doing on behalf of their class what the representatives, hirelings and agents of the bourgeoisie did for the former ruling classes.

Lassalle once said that the German workers of his day—more than half a century ago—were poor in understanding of their own poverty. The revolutionary development of the proletariat consists also in the fact that it arrives at an understanding of its oppressed position, its poverty, and rises against the ruling classes. This gives it the possibility of seizing political power. But the taking of political power essentially reveals to the proletariat for the first time the full picture of its poverty in respect of general and specialized education and government experience. The understanding by the revolutionary class of its own inadequacies is the guarantee that these will be overcome.

It would undoubtedly be most dangerous for the working class if its leading circles were to suppose that with the conquest of power the main thing had been done, and were to allow their revolutionary conscience to go to sleep upon what has been achieved. The proletariat did not, indeed, carry through the revolution in order to make it possible for thousands or even tens of thousands of advanced workers to settle into jobs in the soviets and commissariats. Our revolution will fully justify itself only when every toiling man and woman feels that his or her life has become easier, freer, cleaner and more dignified. This has not yet been achieved. A hard road still lies between us and this, our essential and only goal.

In order that the life of the working millions may become easier, more abundant and richer in content, it is necessary to increase in every sphere the organization

and efficiency of work and to attain an incomparably
higher level of knowledge, a wider horizon for all those
called to be representatives of the working class in all
fields of their activity. While working it is necessary to
learn. It is necessary to learn from all from whom anything can be learnt. It is necessary to attract and draw
in all forces that can be harnessed to work. Once more
—it is necessary to remember that the masses of the
people will evaluate the revolution, in the last analysis,
by its practical results. And they will be quite right in
so doing. Yet there can be no doubt that a section of
Soviet officials have adopted the attitude that the task
of the working class has been fundamentally fulfilled
by the mere calling to power of workers' and peasants'
deputies who cope 'somehow' with their work. The
Soviet régime is the best régime for the workers' revolution just because it most truly reflects the development
of the proletariat, its struggle, its successes, but also its
inadequacies, including those of its leading stratum.
Along with the many thousands of first-class people
whom the proletariat has advanced from its ranks,
people who learn and make progress, and who undoubtedly have a great future before them, there are
also in the leading Soviet organs not a few half-equipped
people who imagine themselves to be know-alls. Complacency, resting content with small successes—this is
the worst feature of Philistinism, which is radically in-
imical to the historical tasks of the proletariat. Nevertheless, this feature is also to be encountered among
those workers who, with more or less justification, can
be called advanced: the heritage of the past, petty-
bourgeois traditions and influences and finally, just the
demand of strained nerves for rest, all do their work.
In addition, there are fairly numerous representatives
of the intelligentsia and semi-intelligentsia who have
sincerely rallied to the cause of the working class but
have not yet had a thorough internal burn-out and so
have retained many qualities and ways of thought which
are characteristic of the petty-bourgeois milieu. These,
the worst elements of the new régime, are striving to
become crystallized as a Soviet bureaucracy.

I said 'the worst' without forgetting the many
thousands of technicians merely lacking in ideas who
are employed by all Soviet institutions. Technicians,
'non-party' specialists, carry out their tasks, well or
badly, without accepting responsibility for the Soviet
régime and without charging our party with responsibility for themselves. It is necessary to make use of

them in every possible way, without demanding from them what they cannot give . . . Our own bureaucrat, however, is real historical ballast—already conservative, sluggish, complacent, unwilling to learn and even expressing enmity to anybody who reminds him of the need to learn.

This is the genuine menace to the cause of communist revolution. These are the genuine accomplices of counter-revolution, even though they are not guilty of any conspiracy. Our factories work not better than those belonging to the bourgeoisie, but worse. The fact that a number of workers stand at their head, as managers, does not in itself solve any problems. If these workers are filled with resolve to achieve great results (and in the majority of cases this is so or will become so), then all difficulties will be overcome. It is necessary to move from all directions towards a more intelligent, more improved organization of the economy and command of the army. It is necessary to arouse initiative, criticism, creative power. It is necessary to give more scope to the great mainspring of emulation. At the same time it is necessary to draw in specialists, to give opportunities to all talents, both those that emerge from the depths and those that remain as a legacy from the bourgeois régime. Only a wretched Soviet bureaucrat, jealous for his new job, and cherishing this job because of the personal privileges it confers and not because of the interests of the workers' revolution, can have an attitude of baseless distrust towards any great expert, oustanding organizer, technician, specialist or scientist—having already decided on his own account that 'me and my mates will get by somehow'.

In our General Staff Academy there are some party comrades now studying who have in practice, in bloody experience, conscientiously understood how hard is the stern art of war and who are now working with the greatest attention under the guidance of professors of the old military school. People who are close to the Academy tell me that the attitude of the pupils to their teachers is not at all determined by political factors, and apparently it is the most conservative of the teachers who is honoured with the most notable marks of attention. These people want to learn. They see beside them others who possess knowledge, and they do not sniff, do not swagger, do not shout, 'tossing their Soviet caps in the air'—they learn diligently and conscientiously from the 'tsarist generals', because these generals know what the communists do not know and what the communists need to know. And I have no doubt that, when they have learnt, our Red military academicians will make substantial corrections to what they are now learning, and perhaps will even make some fresh contributions of their own.

Insufficient knowledge is, of course, not a fault but a misfortune, and moreover a misfortune which can be put right. But this misfortune becomes a fault and even a crime when it is supplemented by complacency, reliance on 'maybe' and 'most likely',[1] and an attitude of envy and hatred towards anybody who knows more than oneself.

You asked why this question of the military specialists has aroused such passion. The essence of the matter is that behind this question, if we dig far enough, two trends are hidden: one, which proceeds from an appreciation of the tasks confronting us, endeavours to utilize all the forces and resources which the proletariat has inherited from capitalism—to rationalize, i.e., to comprehend in practice, all social work, including military work, introducing in every sphere the principle of economy of forces, achieving the greatest possible results with the minimum of sacrifices—really to create conditions under which it will be easier to live. The other trend, which fortunately is much less strong, is nourished by the moods of limited, envious, complacent (and yet at the same time unsure of itself) Philistine-bureaucratic conservatism . . . . 'We're managing somehow, aren't we, so we'll keep on managing all right.' It isn't true! We shall not manage 'somehow' in any case: either we shall manage completely, as we ought, in accordance with science, applying and developing all the powers and resources of technique, or we shall not manage at all, but collapse in ruin. Who has not understood this has not understood anything.

Returning to the question you raise about the military specialists, let me tell you this, from my own direct observation. There are certain corners in our armed forces where 'distrust' of the military specialists is particularly flourishing. What corners are these? The most cultured, the richest in political knowledge of the masses? Not a bit! On the contrary, these are the most deprived corners of our Soviet republic. In one of our armies it was considered not long ago a mark of the highest revolutionariness to jeer rather pettily and

---

[1] Trotsky uses the Russian expressions which traditionally symbolize a lazy-minded, happy-go-lucky attitude to serious problems—Trans.

stupidly at 'military specialists', i.e., at all who had studied in military schools. Yet in this very same army practically no political work was carried on. The attitude there was no less hostile, perhaps more so, towards communist commissars, these political 'specialists', than it was towards the military specialists. Who was sowing this hostility? The worst sort among the new commanders—military half-experts, half-partisans, half-party people who did not want to have anyone around them, be they party workers or serious military workers. These are the worst sort of commanders. They are ignorant but they do not want to learn. Their failures—how could they have successes?— they always seek to explain by somebody else's treachery. They quail miserably before any change in the morale of their units, for they lack any serious moral and military authority. When a unit, not feeling the hand of a firm leader, refuses to attack, they hide behind its back. Hanging on for dear life to their jobs, they hate the mere mention of military studies. For them these are identified with treachery and perfidy. Many of them, getting finally into a hopeless mess, have ended up by simply rebelling against the Soviet power.

In those units where the level of the Red Army men's morale is higher, where political work is carried on, where there are responsible commissars and party cells, they have no fear of the military specialists; on the contrary, they ask for them, use them and learn from them. Moreover in those units they catch the real traitors much more successfully and shoot them in good time. And, what is most important of all, those units win victories.

That is how it is, dear friend. Now, perhaps, you can better grasp the root of the differences that exist on the question of military and other specialists.

Appendix Two

FUNCTIONARISM IN THE ARMY AND ELSEWHERE

December 3, 1923

I.

In the course of the last year, the military workers and I have on many occasions exchanged opinions, orally and in writing, on the negative phenomena visible in the army and stemming from mouldy **functionarism.** I dealt with this question thoroughly enough at the last congress of the political workers of the Army and the Fleet. But it is so serious that it seems to me opportune to speak of it in our general press, all the more so because the malady is in no wise confined to the army.

Functionarism is closely related to bureaucratism. It might even be said that it is one of its manifestations. When, as a result of being habituated to the same form, people cease to think things through; when they smugly employ conventional phrases without reflecting on what they mean; when they give the customary orders without asking if they are rational; when they take fright at every new word, every criticism, every initiative, every sign of independence—that indicates that they have fallen into the toils of the functionarist spirit, dangerous to the highest degree.

At the conference of the military-political workers, I cited as an, at first sight, innocent example of functionary-ideology, some historical sketches of our military units. The publication of these works dealing with the history of our armies, our divisions, our regiments, is a valuable acquisition. It attests that our military units have been constituted in battle and in technical apprenticeship not only from the standpoint of organization but also from the spiritual standpoint, as living organisms; and it indicates the interest shown in their past. But most of these historical outlines—there is no reason to hide the sin—are written in a pompous and bombastic tone.

Even more, certain of these works make you recall the old historical sketches devoted to the Guard Regiments of the Czar. This comparison will no doubt provoke gleeful snickers from the White press. But we would be old washrags indeed if we renounced self-criticism out of fear of providing our enemies with a trump. The advantages of a salutary self-criticism are incomparably superior to the harm that may result for us from the fact that Dan or Chernov will repeat our criticism. Yes, let it be known to the pious (and impious!) old ladies who fall into panic at the first sound of self-criticism (or create panic around themselves).

To be sure, our regiments and our divisions, and with them the country as a whole, have the right to be proud of their victories. But it wasn't only victories that we had, and we did not attain these victories directly, but along very roundabout roads. During the civil war we saw displays of unexampled heroism, all the more worthy because it most often remained anonymous, collective; but we also had cases of weakness, of panic, of pusillanimity, of incompetence and even of treason. The history of every one of our "old" regiments (four or five years is already old age in time of revolution), is extremely interesting and instructive if told truthfully and vibrantly, that is, the way it unfolded on the battlefield and in the barracks. Instead of that, you often find a heroic legend in the most banally functionarist manner. To read it, you would think there are only heroes in our ranks; that every soldier burns with the desire to fight; that the enemy is always superior in numbers; that all our orders are reasonable, appropriate for the occasion; that the execution is brilliant, etc.

To think that by such procedures a military unit can be enhanced in its own eyes, and a happy influence be exerted on the training of the youth, is to be imbued with the mouldy functionarist spirit. In the best of cases, this "history" will leave no impression at all; the Red soldier will read it or listen to it the way his father listened to **Lives of the Saints:** just as magnificent, uplifting, but not true to life. Those who are older or who participated in the civil war, or who are simply more intelligent, will say to themselves, the military people too are throwing sand in our eyes; or simpler yet: they're giving us a lot of hokum. The more naive, those who take everything for good coin, will think: "How am I, a weak mortal, to raise myself to the

level of those heroes? ..." And in this way, this "history," instead
of raising their morale, will depress them.*

Historical truth does not have a purely historical interest for us.
These historical sketches are needed by us in the first place as a
means of **education.** And if, for example, a young commandant
accustoms himself to the conventional lie about the past, he will
speedily reach the point of admitting it in his daily practical and even
military activity. If, for example, he happens to commit a blunder,
he will ask himself: Ought I report this truthfully? He must! But
he has been raised in the functionarist spirit, he does not want to
derogate the heroes whose exploits he has read in the history of his
regiment; or, quite simply, the feeling of responsibility has deadened
in him. In that case, he trims, that is, he distorts the facts, and de-
ceives his superiors. And false reports of subordinates inevitably
produce, in the long run, erroneous orders and dispositions of the
superiors. Finally—and this is the worst thing—the commandant is
simply afraid to report the truth to his chiefs. Functionarism then
assumes its most repulsive character: lying to please superiors.

Supreme heroism, in the military art as in the revolution, is veracity
and the feeling of responsibility. We speak of veracity not from the
standpoint of an abstract morality that teaches that man must never
lie nor deceive his neighbour. These idealistic principles are pure
hypocrisy in a class society where antagonistic interests, struggles
and war exist. The military art in particular necessarily includes
ruse, dissimulation, surprise, deception. But it is one thing consciously and deliberately to deceive the enemy in the name of a
cause for which life itself is given; and another thing to give out
injurious, false information, assurances that "all goes well," out of
false modesty or out of fawning or lick-spittlery, or simply under the
influence of bureaucratic functionarism.

II.

Why do we now deal with the question of functionarism? How
was it posed in the first years of the revolution? We have the army
in mind here too, but the reader will himself make the necessary
analogies in all other fields of our work, for there is a certain parallel
in the development of a class, its party, its state, and its army.

The new cadres of our army were supplemented by revolutionists,
fighting militants, and partisans, who had made the October Revo-

---

* To be sure, not only in the military world, but everywhere else, including
the field of art, there are advocates of the conventional lie which "uplifts the
soul." Criticism and self-criticism seem to them an "acid" that dissolves the
will. The petty bourgeois, as is known, needs pseudo-classical consolation
and cannot bear criticism. But the same cannot hold for us, revolutionary
army and revolutionary party. The youth must relentlessly combat such a
state of mind in their ranks.—L.T.

lution and who had already acquired a certain past and above all a character. The characteristic of these commandants is not lack of initiative or, more exactly, an inadequate understanding of the need of co-ordination in action and of firm discipline ("partisanism"). The first period of military organization is filled with the struggle against all forms of military "independence." The aim then is the establishment of rational relationships and firm discipline. The years of civil war were a hard school in this respect. In the end, the balance necessary between personal independence and the feeling of discipline was successfully established among the best revolutionary commandants from the first levy.

The development of our young army cadres takes place quite differently during the years of truce. As a young man, the future commandant enters the Military School. He has neither revolutionary past nor war experience. He is a neophyte. He does not build up the Red Army as the old generation did; he enters it as a ready-made organization having an internal regime and definite traditions. Here is a clear analogy with the relationships between the young communists and the old guard of the party. That is why the means by which the army's fighting tradition, or the party's revolutionary tradition, is transmitted to the young people is of vast importance. Without a continuous lineage, and consequently, without a tradition, there cannot be stable progress. But tradition is not a rigid canon nor an official manual; it cannot be learned by heart nor accepted as gospel; not everything the old generation says can be believed merely "on its word of honour." On the contrary, the tradition must, so to speak, be conquered by internal travail; it must be worked out by oneself in a critical manner, and in that way assimilated. Otherwise the whole structure will be built on sand. I have already spoken of the representatives of the "old guard" (ordinarily of the second and third order) who inculcate tradition into the youth after the example of Famusov: "Learn by looking at the elders: us, for example, or our deceased uncle. . . ." But neither from the uncle nor from his nephews is there anything worth while learning.

It is incontestable that our old cadres, which have rendered immortal services to the revolution, enjoy very great authority in the eyes of the young military men. And that's excellent, for it assures the indissoluble bond between the higher and lower commands, and their link with the ranks of the soldiers. But on one condition: that the authority of the old does not exterminate the personality of the young, and most certainly that it does not terrorize them.

It is in the army that it is easiest and most tempting to establish this principle: "Keep your mouth shut and don't reason." But in the military field, this "principle" is just as disastrous as in any other. The principal task consists not in preventing but in aiding the young commandant to work out his own opinion, his own will, his personality, in which independence must join with the feeling of discipline. The commandant and, in general any man trained merely to say: "Yes, sir!" is a nobody. Of such people, the old satirist, Saltykov, said: "They keep saying Yes, Yes, Yes, till they get you in a mess." With such yes-men the military administrative apparatus, that is, the totality of military bureaus, may still function, not without some success, at least seemingly. But what an army, a mass fighting organization, needs is not sycophantic functionaries but men who are strongly tempered morally, permeated with a feeling of personal responsibility who, on every important question, will make it their duty to work out conscientiously their personal opinion and will defend it courageously by every means that does not violate rationally (that is, not bureaucratically) understood discipline and unity of action.

The history of the Red Army, like that of its various units, is one of the most important means of mutual understanding and of establishing the link between the old and the new generation of military cadres. That is why bureaucratic lick-spittlery, spurious docility and all other forms of empty well-wishers who know what side their bread is buttered on, cannot be tolerated. What is needed is criticism, checking of facts, independence of thought, the personal elaboration of the present and the future, independence of character, the feeling of responsibility, truth toward oneself and toward one's work. However, those are things that find in functionarism their mortal enemy. Let us therefore sweep it out, smoke it out, and smoke it out of every corner!

INDEX